Leading and Managing Professional Services Firms in the Infrastructure Sector

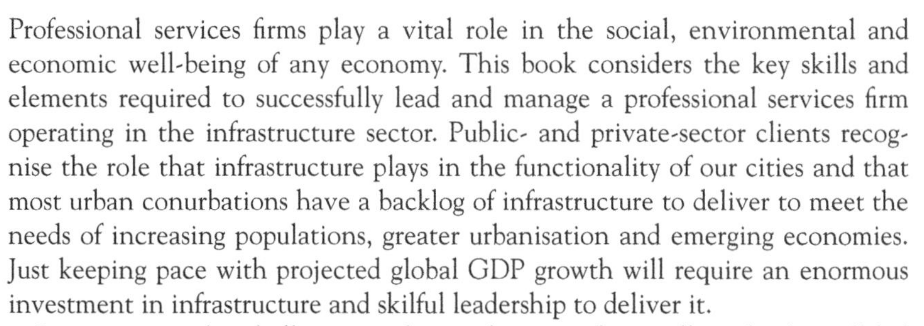

Professional services firms play a vital role in the social, environmental and economic well-being of any economy. This book considers the key skills and elements required to successfully lead and manage a professional services firm operating in the infrastructure sector. Public- and private-sector clients recognise the role that infrastructure plays in the functionality of our cities and that most urban conurbations have a backlog of infrastructure to deliver to meet the needs of increasing populations, greater urbanisation and emerging economies. Just keeping pace with projected global GDP growth will require an enormous investment in infrastructure and skilful leadership to deliver it.

In response to this challenge, professional services firms will need to be well-led and well-managed to be successful and sustainable in the long-term. Such organisations must provide high-value advice, design, knowledge and innovations to get more out of the existing assets and to plan and design new assets with greater integrity and construct them more productively, efficiently and effectively. This book provides practical frameworks for emerging operational managers and future project leaders to prepare them to successfully manage these firms and deliver such projects in the face of new and often disruptive technologies and shifting corporate landscapes.

The book is essential reading for aspiring leaders operating in all infrastructure market sectors including energy, water, sewerage, road, rail, ports, airports, education, health, justice, retail, entertainment, property and development sectors.

Tim Ellis has over thirty-five years' experience in the engineering consulting industry having worked in operational management and project leadership roles across Australia, the Middle East, South East Asia and the UK. He has transitioned from technical specialisation, project management, operational management, project leadership and more recently to risk management. Over that time, Tim has successfully developed a diverse range of professional services firms through organic growth, outsourcing and acquisition. Major infrastructure projects have been a key platform for the growth of these firms. He is a Fellow and Engineering Executive of the Institution of Engineers, Australia.

Leading and Managing Professional Services Firms in the Infrastructure Sector

Tim Ellis

Routledge
Taylor & Francis Group
LONDON AND NEW YORK

First published 2018
by Routledge
2 Park Square, Milton Park, Abingdon, Oxon OX14 4RN

and by Routledge
711 Third Avenue, New York, NY 10017

Routledge is an imprint of the Taylor & Francis Group, an informa business

British Library Cataloguing-in-Publication Data
A catalogue record for this book is available from the British Library

Library of Congress Cataloging-in-Publication Data
Names: Ellis, Tim, 1962- author.
Title: Leading and managing professional services firms in the infrastructure sector / Tim Ellis.
Description: Abingdon, Oxon ; New York, NY : Routledge, 2018. | Includes bibliographical references.
Identifiers: LCCN 2017043742 | ISBN 9780815379171 (hardback : alk. paper) | ISBN 9780815379188 (pbk. : alk. paper) | ISBN 9781351216623 (ebook)
Subjects: LCSH: Consulting firms. | Engineering firms. | Public works. | Public contracts.
Classification: LCC TA216 .E525 2018 | DDC 363.6068--dc23
LC record available at https://lccn.loc.gov/2017043742

ISBN: 978-0-8153-7917-1 (hbk)
ISBN: 978-0-8153-7918-8 (pbk)
ISBN: 978-1-351-21662-3 (ebk)

Typeset in Goudy
by Sunrise Setting Ltd, Brixham, UK

Contents

Figures and tables

Figures

Tables

Foreword

There is widespread acceptance of the importance of infrastructure and the need for more of it to support economic growth, social well-being and environmental sustainability. Global growth is driving an increasing need for vital infrastructure. The World Bank estimates that US$60–70 trillion of investment is needed and only US$45 trillion is available from traditional sources. That means the remaining US$14–25 trillion has to come from collaboration with the private sector. Complexity is now the byword for major infrastructure project delivery.

Infrastructure: a single word that unites all the things that make our modern lives possible. It is a word that connects our cities and towns with all the cities and towns in the world. Roads, railways, cycle paths, airports, wires, pipes, electricity, gas, water, telephone, internet, hospitals, schools, prisons, courts, community centres, sporting facilities, scout halls and the list goes on. These things are all around us – and yet we often barely notice them, unless they impact our daily lifestyles, let alone think about how they got there and who pays for them. Our communities have an insatiable demand for infrastructure. At the same time projects are getting bigger and ever more complex to deliver.

Large megaprojects would appear to pose special problems. Considerable time delays and cost overruns seem to be commonplace. Poor forecasts of the demand for the infrastructure services, leading to severe revenue shortfalls on infrastructure investments have led to descriptions such as 'appraisal optimism' and 'optimism bias' often known as 'deal fever' that are now embedded into our business processes.

If success in megaproject planning and management is identified as projects being delivered on budget, on time and with expected benefits, then the evidence is that approximately one out of ten megaprojects is on budget, one out of ten is on schedule, and one out of ten is delivering on the benefits.

By far the most significant reason for failure is the inadequacy of the business case. Following this aspect, the next most important contributors are environmental impacts, inter-participant disputes, economic conditions, interaction of design with construction and operations and the human factors in our project teams and our organisations. Solutions to these problems are better risk management and greater diligence at the project-definition stage, with the aim of reviewing the project objectives, scope, specifications and definitions detailed in the business case to ensure they are fully comprehensive and address all of the project requirements in the short, medium and long term. Disputes and claims can also be as a result of inadequate specification giving rise to variations and consequently claims, there is clearly a need to put considerable effort up front into ensuring that the business case represents the interests of all project constituents, in terms of the scope of the project and its objectives.

There is much to manage if projects are to be delivered on time, inside budget and deliver the quality and functionality needed to support economic growth and social well-being, while minimising the negative impacts on the environment. To do this effectively requires people, and lots of them. But not just anyone can deliver complex infrastructure projects. It requires deep skills and multidisciplinary teams: engineers; project managers; architects; quantity surveyors; lawyers; accountants; financial and commercial specialists. Successful projects are those where these professionals from very different technical backgrounds come together and effectively collaborate. A high-performance culture needed for success doesn't just magically happen when professionals come together on a project. It needs to be created, nurtured and managed through effective communications, mutual respect and a disciplined approach to the task. It all starts in the professional services firm.

This book can be more accurately described as an operating manual on how to successfully manage these professional services firms that hire, manage and develop the people needed to successfully deliver complex infrastructure projects. Professionals are highly trained and technically proficient individuals, but they are most successful when nurtured and developed; in other words, well-led and well-managed. This book is a welcome addition to the literature that all professionals set on a career in infrastructure would do well to have close at hand.

Darrin Grimsey
Lead Client Service Partner
Transaction and Advisory Services, Government & Public Sector
Infrastructure Advisory
Ernst & Young

Acknowledgements

The author would like to acknowledge the peers and colleagues whom he has collaborated with over thirty-five years in a number of professional services firms and on a number of major projects in developing these insights and perspectives. Also the colleagues who have reviewed this book and provided key insights as to what constitutes successfully leading and managing a professional services firm in the infrastructure sector.

Acronyms and glossary

ABC	Anti-bribery and Corruption or activator-behaviour-consequence analysis
AC	Actual Costs
Alliance	Collaborating form of project delivery, usually with the Owner, Designer, Contractor and Operator
BIM	Building Information Management, an information management system not just for buildings
BMS	Business Management System, housing the organisational business processes
Certificate	A statement of conformance that the design complies with the OPR
CD	Certified Design (aka Stage 2, design for approval with the reviewers for compliance with the OPR), broadly 85% design completion
CJV	Construction Joint Venture, between allied construction organisations
CPI	Cost Performance Index, part of an earned value measurement
CPS	Construction Phase Services, to support a D&C Contractor during the construction phase
CRM	Client Relationship Management, a business process for managing relationships with preferred clients
D&C	Design & Construct, form of project delivery
DCN	Design Change Notice, pursuant to the contract
DCR	Design Change Register, summarising the DCNs
DJV	Design Joint Venture, between allied design partners
DMP	Design Management Plan (aka Design & Engineering Management Plan, Project Execution Plan)
DoA	Delegation of Authority, from the project governors to the PMT
DPR	Design Package Register, of design packages based on the WBS with target dates at each design stage
DSA	Design Services Agreement (aka Consultant or Professional Services Agreement)
DSO	Days of Sales Outstanding

EAC	Estimate at Completion, part of an earned value assessment
EBIT	Earnings Before Interest and Tax
ECI	Early Contractor Involvement, form of project delivery
EOI	Expression of Interest
EOT	Extension of Time, pursuant to the contract
EPC	Engineering, Procure & Construct, form of project delivery
EPCM	Engineering, Procurement and Construction Management
EQ	Emotional Qualities or Quotient
ETC	Estimate to Complete, part of an earned value assessment
EV	Earned Value
EVM	Earned Value Measurement
FM	Facilities Management, form of delivery for the O&M phase
GFC	Global Financial Crisis
GNG	Go No Go, to proceed with pursuing or submitting a proposal
HVHR	High Value, High Risk tenders or projects
IA	Independent Assurance, of a tender or project, incorporating people, process and quality assurance
IFC	Issued For Construction (aka For Construction Documentation), 100% design completion
JVA	Joint Venture Agreement, between allied design partners
KBI	Key Behavioural Indicator, measure of the behaviours that create the desired culture
KPI	Key Performance Indicator, measure of performance
KRA	Key Result Areas, describing the project objectives
KSEA	Knowledge, Skills, Experience & Attributes of personnel
Milestone	A contractual date that triggers an EOT or delay costs
NPS	Net Promoter Score, used for soliciting client feedback
NTA	Net Tangible Assets
O&M	Operations & Maintenance phase of the project
OPR	Owner's Project Requirements (aka Principal's Project Requirements, Project Scope & Technical Requirements)
Owner	Project Owner (aka Principal, End Client, Asset Owner)
PD	Preliminary Design (aka Stage 1, schematic design), broadly 35% design completion
PM	Project Manager
PMC	Project Management Consultant, responsible for managing the client's project
PMO	Project Management Office, including template management plans, project controls, procedures and tools
PMT	Project Management Team (aka Design Management Team)
PPP	Public Private Partnerships, or P3, form of project delivery
PRP	Peer Review Panel, an independent panel of highly experienced practitioners
PV	Planned Value

RACI	Responsibility, Accountability, Consulted and Informed matrix for project interactions
Reviewers	Third-party reviewers including Owner's Engineer or Technical Advisor, Independent Reviewer, Verifier or Certifier, Proof Engineer, Building Certifier, Approving Authorities, Utility/Asset Owners
RFI	Request for Information, form a construction contractor fills in for further information or technical query
RFP	Request for Proposal
R&O	Risks and Opportunities
ROI	Return on Investment on non-project time activities
SFAIRP	So Far As Is Reasonably Practical, a risk-based approach to design solutions
SID	Safety in Design, a process to design out construction and operational risks
SPI	Schedule Performance Index, part of an earned value measurement
SQ	Spiritual or Social Qualities or Quotient
STEM	Science, Technology, Engineering and Mathematics
Target date	A target date in the DPR that is not a contractual Milestone
TAN	Tender Advice Notice, deliverable of a tender design for a D&C contractor
T&E	Time and Expense (aka Time and Materials) for a cost reimbursement contract
TSA	Tender Services Agreement
TSMP	Tender Services Management Plan
TQ	Technical Qualities or Quotient
VE	Value Engineering (aka Value Management, options analysis)
VfM	Value for Money
VN	Variation Notice, pursuant to the contract
WBS	Work Breakdown Structure
WHS	Workplace Health and Safety, legislation in each jurisdiction
WIP	Work in Progress
WPT	Wider Project Team

1 Introduction

Professional services firms play a vital role in the social, environmental and economic well-being of any economy. This book provides the key success factors for leading and managing a professional services firm operating in the infrastructure sectors. Public- and private-sector clients recognise the role that infrastructure plays in the liveability of our cities and most urban conurbations will have a backlog of infrastructure representing an enormous challenge with increasing populations, greater urbanisation and emerging economies. Just keeping pace with projected global GDP growth will require an enormous investment in infrastructure (Grimsey, 2017).

Consequently, professional services firms are forecast to be high-growth firms of the future. In response to this challenge, professional services firms will need to be well-led and well-managed to be successful and sustainable in the long-term. Professional services firms provide high-value advice, design, knowledge and innovations to get more out of the existing assets and to plan and design new assets with greater design integrity and to construct them more productively, efficiently and effectively. With the advent of new and often disruptive technologies, greater complexity in shareholders' needs and wants and more demanding shareholders, future success will not be assured by back-casting past practices and behaviours. This book provides practical frameworks for emerging operational managers and future project leaders to climb the learning curve for successfully managing these firms of the future.

The purpose of this book is to leverage the learnings from the last three decades in professional services firms, serving both the private and public infrastructure sectors. In this context professional services mean technical services in architecture, engineering, commercial, legal and any technical advisory services. This book is a compilation of discussion papers, thought pieces and insights from the early 1980s through the peaks and troughs of these market sectors as firms have transitioned from partnerships to large international businesses.

With over thirty-five years of diverse experiences, the author has been a highly experienced leader of consulting and engineering firms with unparalleled experience in the planning, design and delivery of major building and infrastructure projects. Over this time he has developed unique insights into the challenges related to managing a diverse portfolio of projects from consulting, design, project management, to construction services. The learnings presented in this book are evidence-based with industry references. An extensive list of references is provided for a deeper dive into any of the chapters.

While many of these insights are based on Australian experiences and market conditions, the learnings are applicable to all geographies and market sectors as the Australian marketplace exhibits:

- Multi-sector projects ranging from mega oil & gas and resources projects, to small advisory commissions to government agencies
- A relatively small market by international comparisons, diverse range of technical services disciplines within the one firm
- A high percentage of public private partnership (PPP or P3) projects in the delivery of social and transport infrastructure
- A high percentage of design and construct (D&C) form of project delivery along with alliance contracting, early contractor involvement and other fast-track and accelerated forms of project procurement
- Often onerous risk allocation from Project Owners.

This book is written in the collective sense as a trusted partner, coach or advisor sitting alongside the reader in a workshop or planning session. It is based on five key tenets:

Strategic Approach: A framework for 'blue-sky' thinking for developing emerging businesses, the winning of strategic projects, against the odds and then successful delivery.

Operational Management: Insights into the management of a diverse range of professional services from strategic planning, architecture, project management, all the engineering disciplines to technical services across the civil- and social-infrastructure market sectors.

Project Leadership: Appreciation of the attributes of project leadership based on balancing strategy and governance roles for the teaming, tendering and delivery phases of large and complex major infrastructure projects. This also involves the turn-around of distressed projects, incorporating hard-earned lessons learned.

Risk Management and Governance: Harnessing learnings from all aspects across the project life cycle mitigating risks and harnessing opportunities. The project leaders of the future will be prudent risk takers providing assurance of approach to their project and corporate governors.

Collaborative Relationships: Understanding of the importance of highly collaborative relationships with Project Owners, allied design partners, contractors and related stakeholders. Knowledge that management of diverse professional services firms or leadership of complex major projects requires strong interpersonal skills developing sound relationships based on trust and respect with key people and stakeholders. This forms the basis of developing an organisational or project culture of high-performance.

The application of these five fundamental principles demonstrates that success is achieved through the achievements of a highly collaborative operational management or project leadership team.

Tim Ellis
Melbourne, Australia
September 2017

2 Strategy

> Great minds discuss ideas, average minds discuss events, small minds discuss people.
>
> (Eleanor Roosevelt)

This chapter begins by setting the scene as to what a professional services firm of the future would look like in the future, the 'flag on the hill' so to speak and the strategic planning to achieve that desired end-state. The term strategy is used in the context of *The Art of War* (Tzu, 1957) as looking at the long-game and considering the influencers on the future success of the firm. It looks at some of the externalities of market drivers and changes, suggests a process for strategic planning, how to establish a vision for success, concluding with an indicative organisational strategy map.

2.1 Market outlook

With emerging economies, increasing population growth and greater urbanisation, there is a growing demand for public infrastructure to sustain our communities' well-being and sustainability. There is widespread acceptance that public infrastructure seeds economic development; however, often the supply lags behind the demand, given short-term political drivers and funding. Therein the private sector is taking on greater participation in the provision of public infrastructure. There will be an insatiable demand for faster and more reliable transport infrastructure capturing the value of travel time saved, increased density of urban places and communities and value capture for employment and social well-being, extension of existing infrastructure asset life and asset recycling.

Local and international market conditions are highly uncertain, providing greater uncertainty in the project pipeline. Large infrastructure projects have a long gestation and have a history of time delays and cost overruns. There is increasing risk transfer from Project Owners, often with very onerous risk allocation, often not consistent with Abrahamson's basic principles of risk allocation to the party best able to manage and control the risk (Abrahamson, 2017). These clients are seeking to transfer greater risks to downstream suppliers, many of these not being covered by insurance, commonly referred to as 'balance sheet' risks. Projects are also becoming more complex with often diverse and eclectic stakeholders and third parties. Also many clients are allocating delivery risk to Design and Construct (D&C) contractors, Facilities Management (FM) contractors or Project Managers. This pass-through risk allocation to downstream consultants is often not equitable, placing greater emphasis on heightened risk management and assurance.

So considering the demand and supply sides of the economic equation, the requirements for the professional services firm of the future are likely to be far more challenging. Professional services project leaders and operational mangers of the future will need to have far more developed knowledge, skills, experience and attributes (KSEA) than the past. Technical excellence alone is no longer the key ingredient to success. The core attributes of new graduates will be problem solving, having an ability to collaborate and influence others with the commitment and resilience to respond to the project challenges and therefore achieve desired outcomes.

2.2 The firm of the future

> Life is short, art long, opportunity fleeting, experience misleading, and judgement difficult.
>
> (Hippocrates)

It is imperative that the firm articulates what success would look like in the future to engender emotional engagement, commitment and loyalty from all staff. The following is an example of a vision of success.

Longevity is a very powerful brand statement. It not only recognises a portfolio of outstanding projects, being responsive to the changing market conditions and a statement of success, but also it emotes a longer term perspective and a sense of future success. Many professional services firms are launching from a solid foundation of the initial founders and partners (Maister, 1993). However, past success is not a predictor of future success and this chapter provides examples so that professional services can continue to evolve and not become stuck in the legacies of the past.

While we can derive a series of key performance indicators to measure success, this section sets out a vision of what operating a successful professional services firm would feel like, the pulse, the mojo, or the organisational DNA.

Using the cross-stitch of **SUCCESS**, the following is a representation of what success would look like for an inspirational leader of a professional services firm. This is just one example of how an operational manager may impart what is important to them, motivates them and what they care about and value. Like an inspirational 'sermon from the mount' of religious leaders, this narrative engages their people with a common sense of purpose and direction. Later in the book, we will discuss how this higher sense of purpose is a key motivator for developing high-performing teams.

Sustainability
Unrelenting
Consulting
Culture
Essence
Success
Special

The first S is for Sustainability

A sustainable firm has scale, size and substance leading to outstanding success. It has a sense of success, it just feels right. A successful firm has the majority of commissions sole sourced or with negotiated contracts. Our preferred clients continue to engage the firm as their consultant of choice, their trusted partner.

A successful professional services firm is growing at a faster rate than the economy, other competitors and is achieving industry profitability to fund its growth for the longer term. A successful firm balances ongoing profitability with long-term growth and diversity in new markets, business streams, client types, new technologies and geographies.

U is for Unrelenting, uncompromising and persistent

In a competitive, volatile and dynamic marketplace, the successful professional services firm will be fiercely competitive and unrelenting in its pursuit for success. It has uncompromising integrity in safety, ethics and relationships and a commitment to technical quality. It will be highly precious of its technical reputation and recognises this reputation could be ruined with just one small project incident or stuff-up. It will exhibit a culture of accountability with zero tolerance for breach of the firm's values and an unwavering compliance to assure technical quality.

As Calvin Coolidge stated:

> Nothing in this world can take the place of persistence. Talent will not. Nothing is more common than unsuccessful men with talent. Genius will not. Unrewarded genius is almost a proverb. Education will not. The world is full of educated derelicts. Persistence and determination alone are omnipotent.

The key personnel will assume full accountability for their actions and inactions. It is often what we fail to do and lack of timely intervention that is our undoing. We will not compromise the values and principles of the firm. We will have a prudent appetite for risk and not shy away, but rather embrace risk and back our key personnel that have the knowledge, skills, experience and attributes (KSEA) to manage a successful outcome.

The first C is for Consulting acumen

The successful professional services firm levers its technical skills bench-strength with other professional services. It embraces the new fields in all aspects, not just technically. It fully embraces and is inclusive of a diverse knowledge-base of new generations, ageism, internationalism, gender balance, workplace flexibility, personal drivers and experiences.

The successful firm balances the mix of higher value advisory services with traditional services throughout the project life-cycle and values each on the client's perception of the value we bring.

The successful firm understands the 'art of consulting', to act as the trusted partner with our preferred clients. We will be client-centric, with confidence, assuredness and self-esteem, to balance the interests of clients, corporate management, shareholders, and a diverse staff mix. Everyone profits, from making a profit.

Our offer will be an invitation to share with the client a sensational experience. We will have a culture of outstanding service delivery. We will focus on creating a sensational experience for our preferred clients. We understand that client loyalty is derived from how we work with our clients, as well as what we deliver.

While we strive for the highest standard of technical excellence, client loyalty will be derived from technical excellence and outstanding service delivery. This will be the very essence of our attitude – we invest in client relationships; we share and collaborate; we put ourselves in our client's shoes; and we are focussed on client care and value.

To earn this trusting relationship with our preferred clients, we will be willing to go first, give a favour to get a favour. As a service provider, our clients will visibly perceive that we are willing to be the first to make an investment in the relationship. We will demonstrate through our applications, not assert, illustrate through our behaviours, not sell.

Our trust-based relationships will be based on open, robust dialogue, ownership and accountability and a focus on results that define the essence of the relationship with our preferred clients. Clients will recognise the value we bring and reward us accordingly.

The second C is for Culture

We will understand that changing our behaviours is the first building block in creating a unique client experience. This change in mindset is part of the high-performance strategy and is underpinned by our values and principles. This is our powerful brand statement.

We will strive for performance that is far beyond the traditional business-as-usual by harnessing the positive energy with a united project team, creating a high-performing team spirit. We will respond to the principal challenge for our project teams to develop a united team culture that will break through traditional barriers and paradigms achieving 'game-changing' outcomes.

This culture of innovation and challenge will yield solutions that are regarded as possible, but the team doesn't immediately know how to achieve them. It will require a new way of thinking and we need to be committed to make it happen. Maintaining the status quo and not challenging existing paradigms and 'sacred cows' is not an option.

We recognise the need for significant mindset change to develop a high-performance leadership, management and project teams that are all 'on-board the bus'. Breakthrough performance is less about individual team members and more about galvanising people around clear, seemingly improbable challenges.

We recognise that our future project leaders will operate in environments quite different from those of yesterday and today. Our leaders and rising stars will be the emotionally intelligent leaders, balancing hard and soft, push and pull tactics. We will be unwavering in our commitment to the hard-coded project delivery functions of scope, time, cost and quality. However, these are outcomes of enabling the true potential of our future leaders through developing self-confidence, interpersonal skills, empathy, self-awareness and personal energy. We will focus on what is important rather than the urgent tasks and balance a best-for-project or best-for-firm approach with personal interests.

We recognise that stakeholders are becoming more demanding, combined with resources constraints and the war for talent, that service delivery will be the key differentiator. We will have the courage to embrace risks and seize opportunities in a planned and managed manner.

E is for Essence

The successful professional services firm has at its core the very essence of its being. These will be the shared value and principles. Everyone will be 'on-the-bus', so to speak. There will be open communication such that everyone is confident they're on the right bus, travelling at the right speed and heading for the right destination.

It will exude an aura of energy, passion, commitment and drive. We will be relentless in our mental toughness, emotionally resistant, flexible, positive and persistent.

The essence of the firm will be to remain focussed, resolved and confident, even when challenges arise. And upon success, there will not be excessive hubris, showing humility, yet celebrating and sharing in the team success.

The essence of our business will be our 'vibe', to be challenged, stretch ourselves, be thrilled in what we achieve and have fun. This will be our 'positive Qi'.

The first S is for Strategic outlook

The sustainable professional services firm will be strategic in its approach. From intimate market scanning of political, economic, social, technological, environmental and legislative (PESTEL) changes, we will pre-empt market changes. We will be responsive in determining our service offering within the service life cycle.

Why do we need to develop, where do we want or need to be, what do we need to do, how will we get there? While the strategic decisions of today will provide the foundations for the future, we recognise our strategies need to be flexible and adaptive to the changing dynamic market given ever-evolving risks and opportunities.

The strategic firm has a balance between the short, medium and longer term horizons. It is very much focussed on consistent sales revenue and operating profit, it plans for growth of the business through offering new services to existing clients and existing services to new clients. It also harnesses the intellectual horsepower of the firm with external advisers to consider new frontiers, both on market sectors, client type, service offerings and geographies.

A global firm is more than an amalgam of international offices. We will be a truly global business with a global outlook and local delivery. We will embrace global sector leaders to drive the business forward working in harmony with local management. Collaboration and knowledge sharing will be first nature, with no barriers to best-for-firm choices.

As a global firm it will have the business systems and processes that support the delivery of projects and add value to provision of these services while mitigating risks.

The final S is for Special and sensational

The firm will be very special. It will be a sensational journey for staff and an awesome experience. Staff will be escalating their career progression at a faster rate to achieve their true potential. We will be a magnet for attracting staff from other firms and retention will be unquestionable. Our future leaders will be bred from our rising stars complemented by strategic new recruits. Acquisitions will be welcomed into our family, not as adopted children, but as integral members of the firm's community. They will not only be a source of technical expertise and new client base, but also a source of talent for the future.

We will be a professional services firm that demonstrates best practice in all facets. That is our people, processes, deliverables and financial outcomes. We recognise that the route to success will have its highs and lows. However, we will relish the journey, learn from the experiences and be relentless in our pursuit for success.

Our journey will be simply sensational.

This example 'sermon from the mount' may at first blush appear to be 'way out of left field' for a technical services firm. However, professional services firms are people businesses and people are emotional and social creatures and they need to and want to know what makes a firm tick. Future leaders should develop their own sermons, based on what they feel and believe in. These are our spiritual qualities and these are discussed further in the latter chapters.

2.3 Strategic planning

In this context, strategic planning means the structured planning process for developing the strategic direction of the firm. Looking at the external factors, the infrastructure challenge can be summarised as:

- Address demands in high population growth areas
- Respond to increasing pressures on existing infrastructure

- Provide better places for communities to come together
- Address greater demand on the justice system with immigration and cultural integration
- Address greater demand on the health system with an aging population
- Provide access to high-quality education infrastructure to support lifelong learning
- Meet growing demand for increased urbanisation and economic activity
- Improve accessibility to employment centres with improved transport systems
- Improve efficiency of freight supply chains
- Manage water security
- Manage pressures on landfills and waste-recovery facilities
- Preserve natural environments and minimise biosecurity loss
- Improve the health of local waterways, flood mitigations and coastal areas
- Transition to a lower carbon-energy economy
- Improve the resilience of critical infrastructure
- Embrace new technologies for sustaining existing infrastructure.

In most geographies, the professional services market is a dynamic marketplace and it is important to have a strategic outlook to market sectors as they progress through their product lifecycle (Product Life Cycle Stages, 2017). Many technical disciplines have passed their most productive phase. By example, the commoditisation and standardisation of urban bridge engineering over recent years. Conversely, there are many emerging trends and new technologies.

An indicative strategy map of a professional services firm is outlined below. This is not intended to be the answer to the question of where the firm will be in three years' time. It is intended to demonstrate the process to be undertaken to considering the causation of the changes to our market sectors (demand) to align our technical capability (supply).

Feedback anecdotally and from staff surveys highlights the need for all staff to understand the 'big picture' (Knight, 2003), where the firm is heading and what the firm may be like in a three- to five-year horizon. Strategic planning is therefore a critical element of leading and managing a professional services firm and its impact on staff, shareholders and clients should not be underestimated.

A strategic planning process will normally consider the political, economic, social, technological, environmental and legislative drivers (PESTEL) approach to identify the cause and effect drivers in each market sector (PESTEL Analysis, 2017). Table 2.1 provides an indicative strategic planning map and what could be a key driver, change or disruption to the future success of the firm.

The McKinsey's three horizons of growth model (McKinsey & Company, 2017a) also provides a useful tool, summarised as follows.

Horizon 1: maintain and defend core business

Most of our immediate revenue-making activity will sit in horizon 1. Our goals in horizon 1 will be mostly around improving margins, bettering existing processes and keeping cash coming in.

Table 2.1 Indicative strategic planning map

Sector	*Political*	*Economic*	*Societal*	*Technological*	*Environmental*	*Legislative*
Energy	Unconventional gas	Cost of power	Exporting coal to depressed markets	New technologies to provide cleaner energy	Green energy developments Reliance on coal-fired power stations	2030 emission targets
Water & sewerage	Drought proof	Agriculture	Water re-use	New technologies in water and waste-water treatment	Replace aging pipelines and assets	
Roads	Accessibility	Better use of existing assets User charging Freight task	Support de-centralised living and work–life balance	Automated vehicles for commuters and freight	Tunnels to be adopted to cross high-density inner urban	Market-led proposals
Rail	Public transport	Transit-oriented developments	Primary mode of transport Value of travel time	Smart phones		Fast rail to provincial cities
Ports	Efficient exports and imports	Sustain Melbourne as the number one port in Australia Second container port	Regeneration of port land	Improvements in freight logistics (just in time) Inland freight terminals		Privatisation
Airports	Tourism	Development to match planned passenger growth	Rail link to the airport	Passenger experience	Carbon emissions	Immigration, border control, security
Education	Smart cities	Growth of the professional services sector	Promote Melbourne as the education hub	Online learning		Privatisation

(Continued)

Table 2.1 (Continued)

Sector	*Political*	*Economic*	*Societal*	*Technological*	*Environmental*	*Legislative*
Health & aged care	Research hubs		Aging population	Smart hospitals Online diagnosis		
Justice	Law and order	Cost of incarceration	Justice matches the crime	Home detention		Judicial penalties
Retail & entertainment			Retail hubs On-demand entertainment	Online selling		
Property	Liveability	Value capture Land use planning	Higher density in inner urban areas transport orientated developments	National broadband network and working remotely	Urban regener-ation of existing industrial land Green buildings	Re-zoning and planning regulations
Key issues	Independent and apolitical	Backlog of critical infrastructure Global infrastructure hub Whole of lifecycle costs	Increasing urbanisa-tion with dependence on service economy Developing regional centres	Digital and disruptive technologies for all stages of the infra-structure lifecycle	Natural disasters Climate change Greenhouse gas emissions	Funding by the private and public sector

Horizon 2: nurture emerging business

Leveraging our existing capabilities and extending it into new areas of revenue-driving activity. There may be an initial cost associated with horizon 2 activities, but these investments should return fairly reliably based on them being an extension of our current proven business model.

Horizon 3: create genuinely new business

Introducing entirely new elements to our business that don't exist today. These ideas may be unproven and potentially unprofitable for a significant period of time.

Many professional services firms unleash the spirit of innovation and have a positive effect on people's everyday lives. However, in many developed countries, there is a shortage of people undertaking the science, technology, engineering and mathematics (STEM) courses at high schools and universities. Also, there are many people retiring and exiting the industry. While the demand is increasing, the supply is decreasing. Part of the reasons for this skills shortage is the lack of role models for aspiring professional services people. While skilled migration and a global mobility might help with smoothing peaks and troughs, there is a need to attract and retain the next generation of highly skilled people.

So, the purpose for creating an inspirational vision of success is to attract and retain new graduates and emerging professionals to not only be part of this industry, but also to be an integral part of the future of the firm.

2.4 Vision, mission and values

Investigating the peer group of large international professional services firms in the legal, accounting and technical professions, there is a close commonality of any vision, mission or value statements.

In today's market, professional services firms in the infrastructure sector are often a diverse group of technical people including planners, architects, engineers, economists, scientists, technicians, consultants and academics. We work together to provide the most effective and sustainable solutions for our clients. Our purpose is simply to:

Help build a better world

Our core values support this aspiration and create an environment where our people can flourish.

A Mission statement should define why we exist, and our vision to clearly identify what we aspire to become. Our core values are the priorities we set for our practices and daily activities and provide our moral or ethical compass, internal 'sat nav'. For example:

Mission

To provide exceptional solutions for our clients in the infrastructure sector through the application of our outstanding technical expertise.

Vision

To help build a better world with the supporting infrastructure.

Values

Innovative: We look ahead, anticipating and responding to change with agility. We are problem-solvers who explore new ideas and we are driven to find the value-for-money solution. We challenge the status quo, think outside of the box, and learn from our experiences.

Integrity: All our relationships are built on trust. We are safe, transparent and ethical. We act with integrity, keep our word, and treat everyone with respect. We empower others and we feel empowered.

Collaborative: Our strength is in the power of our collaboration and teamwork with our clients and partners. We leverage our best-in-class skills, our best practices and our ideas locally and across the globe.

Excellence: We believe in providing high-quality technical advice and value in everything we do, to our clients and to our communities. We seek to provide best-in-class expertise and experiences and informed initiative and risk taking.

Sustainability: This is reflected in our long-term outlook and our hard-earned reputation. We are socially responsible and have a positive impact on our communities. We ensure that sustainable principles are woven into all that we do.

When working in collaborative contracts with our clients and partners, it is appropriate to align the values from each of the participants, to ensure a consistent value-set for the project or programme. Later in this book, we will discuss **values-based assurance** as being the application of these values to guide the organisation and project personnel in their decision making and knowing how to behave in times of uncertainty.

2.5 Organisational Strategy Map

Given our appreciation of the external drivers, the nature of our firm and our vision for the future, an organisational strategy map is then developed. A typical strategy map (Norton, 2004) for a professional services firm is outlined below. Similar to any corporate edict, it is not just the outcome or 'sermon from the mount' that is infectious, but also the process used for determining the strategy and the manner in which it is communicated and kept alive. We know that most strategies fail, not because the strategy itself was flawed, but by its lack of implementation (Collins, 2001).

It is therefore imperative that the appropriate personnel (not just line management) are engaged in the development of the strategy and its cascading through

Table 2.2 Indicative organisational strategy map

Vision *To help build a better world with the supporting infrastructure*			
Financial perspective	1. Sustainability A long-term view, ensuring the strategic decisions made today provide the solid foundation to be the leading professional services firm for years to come. KPI: Quarterly audit index	2. Profitability To achieve profitability targets to generate cash flow to fund our return to shareholders and investment in the future of our people, new technologies and market growth. KPI: Profit targets	3. Growth In targeted market sectors through strategic recruitment and acquisition to achieve sustainable market presence. KPI: Backlog and weighted pipeline targets
Client perspective	4. Service Delivery To deliver the services that exceed the expectations of our clients through break-through performance. To honour the promise made in our offer to deliver the scope of services, on time, to the agreed fee, and to the industry standard. KPI: Trade-ups and client survey index	5. Clients of Choice Determining the clients we wish to work with yielding over two-thirds of our work sole-sourced or procured through negotiated contracts; i.e. to be the consultant of choice to our clients of choice. KPI: Revenue from key client targets	6. Technical Excellence To provide leading-edge skills that deliver technical and commercial outcomes though innovation that represents best value for the client, and being renowned as best in the business. KPI: Rework and claims targets
Process perspective	7. Adaptive Management In uncertain times, to develop a flexible and adaptive organisational arrangement that responds to market conditions and brand client's requirements, through effective operational management. KPI: Margin and utilisation targets	8. Systems and Processes As a global firm ensuring the business systems and processes support the delivery of our projects through effective planning & managing and adding value to the provision of these services while mitigating risks through flawless execution. KPI: Cost of selling and managing targets	9. Project Delivery To instill common project management practices tailored for the scale and complexity of the projects. Provide the project management, controls, leadership, governance and oversight during the critical phases. KPI: Bid and outturn margin targets
People perspective	10. Culture To provide a performance-based culture based on shared and open values that ensure we maintain the soul of small teams with the strength of a large global firm. KPI: Productivity targets	11. People Supporting our human capital through proactive learning and development in technology and consulting skills with effective goal setting and dialogue to achieve their true potential; i.e. to engage the employee of choice with the employer of choice. KPI: Staff retention targets	12. Personal and Professional Development And to ensure that we enjoy our work and share in each other's successes. To ensure everyone achieves their true potential and escalates up their career-development curve. KPI: Staff survey index

the organisation with the same degree of passion and commitment. It is to be expected that this message will be filtered through the line management, based on their own preferences and biases. If the CEO can instil the strategy personally through face-to-face town hall sessions, the intranet and other social media, the greater the chance that everyone will embrace the strategic direction of the firm.

A strategy map is normally based on four core perspectives:

- Financial or outcome perspective
- Client perspective
- Process perspective
- People or stakeholders perspective.

An indicative organisational strategy map is outlined in Table 2.2 using these four perspectives and a dozen key measurable outcomes or key performance indicators (KPI). To keep the strategy accountable, these KPIs should be measured quarterly and the more objective the measurement the more likely the targets will be achieved. What gets measured gets managed (Collins, 2001).

A balanced scorecard strategy map becomes the guiding charter for all personnel and related stakeholders. It can be refreshed periodically in response to market changes and to make sure it is current and relevant to emerging professionals to drive the growth of the firm.

3 The nature of the firm

Having set a course for future success, this chapter considers how to balance a consulting firm with a design firm and a projects-oriented firm. Understanding the different nature of these commissions helps establish the strategy and culture of the firm.

A consulting firm will generally be characterised as providing strategic and technical advisory services to clients with a multitude of small to medium-sized assignments. The key deliverable is often a report. The approach is to 'farm' clients and manage these assignments on a programme, portfolio or on a framework basis with the preferred clients. Assignments are often won and delivered on a 'seller–doer' basis. Given the short-term nature of these assignments, consulting firms generally have moderate utilisation and higher charge rates.

A design firm is characterised by its technical creativity and expertise in the built environment. It balances the built form with functionality. Selection is often through a design competition. These firms desire to shape a better world by giving clients smart design ideas with a social purpose, which will have a positive influence for current and future generations. This built form will integrate with the local and indigenous communities with a unique identity. A design firm is often known for its portfolio of works in the respective social-infrastructure market sector such as rail, aviation, education, health, justice, entertainment, commercial and residential.

A projects-oriented firm generally 'hunts' strategic or major projects, often identified in the proposal pipeline of opportunities well-ahead of the project coming to market. A large project will generally yield higher utilisations and lower rates (assuming contingency is included in the contingent hours not the rates), given greater use of technical staff, utilising a transient workforce and work-packaging to design partners and lower cost operations. The mix of employees to other staff should be based on the cyclic nature of the projects. They also have a very different risk profile and are subject to the peaks and troughs of the market. These projects generally have long gestation periods. While utilisation will be measured for the project duration, it should also consider the bid effort and utilisation of key personnel during the subsequent down-time.

A project-services oriented firm, undertaking roles such as Project Management/Construction Management (PMCM), Project Management Consultant

(PMC) or Engineering, Procurement and Construction Management (EPCM) generally partners with Owners to help them deliver their project, without self-performing the design functions. These projects are characterised by a highly experienced project management team of senior professionals, generally not short of hubris, ego and self-confidence.

Therefore there is a different skill set required for operational management of the consulting operations and advisory services, creative design, design of major projects and project services and this chapter outlines these differences. If the operational manager is required to manage a diverse portfolio of projects, then they will need highly adaptive management skills.

3.1 Consulting and advisory-services oriented firm

Most professional services firms aspire to be further up the 'food chain' providing higher value services to their key clients. As the firm aspires to undertake a greater level of strategic and technical advisory services, the purpose of the section is to provide an overview of the risk aspects compared to the traditional design functions.

Whereas the design processes are generally well-defined, codified and are variations on previous practices, advisory projects are characterised by an ill-defined process and outcomes. Advisory projects are normally secured by developing a unique and compelling methodology of the process to be undertaken to achieve the client's needs (or wants) as either implied or expressed in the request for proposal.

The process is akin to 'baking a cake without a recipe'. The proposal should include the proposed work plan outlining a series of tasks or activities to be undertaken in a sequential order:

- Purpose: What is the purpose for undertaking this task?
- Inputs: Is the process contingent on inputs from the client or third parties?
- Description: How is the process to be undertaken? What are the interfaces, milestones?
- Timing: When is the task required to be completed?
- Responsibility: Who is responsible for undertaking the works and the deliverable and who is responsible for the verification?
- Deliverable: What is the form of the output?

It is recommended that the client should be integrally involved in the process and included in all assumptions, decisions and determinations. This should be undertaken in the spirit of sharing the risk with the client, should the final deliverable not meet the client's or key stakeholder's expectations. As discussed later in the book, the report is for the client's express purposes and should not be issued to and relied upon by third parties.

Adapting the 4Ps of marketing (Kotler, 2012) of people, process, product and price, advisory projects are generally won with the right key people and a compelling process, approach or methodology. If we get these two elements right, the output product and price will also be right. Developing a compelling methodology in the proposal phase can be quite time consuming and often the cost of selling is proportionally higher.

Obviously one of the key KPIs when managing a portfolio or a panel contract is repeat business. Business development activities should be undertaken during the delivery phase and then the need to tender for future commission should be minimised.

The benefits of managing a portfolio of consulting assignments is that risks may be statistically amortised over the portfolio. The business development (aka sales) people and operations personnel are therefore fully integrated.

3.2 Design-oriented firm

Today's cities include large, iconic, complex, long-term and expensive physical and operational, social and economic infrastructures (Moore, 2017). However, they also include small, simple, short-term and inexpensive elements and activities, many of which emerge from informal or quasi-formal community-led initiatives. They can also include unsolicited market-led proposals from the private sector. A design firm understands the local environment and social drivers and is able to balance form with functionality.

Architectural and specialist design firms range from highly creative artistic practices to firms that intimately understand the user's requirements in their specialist market sector such as rail, aviation, education, health, justice, defence, entertainment, commercial and residential.

Often these firms have highly creative personnel and it can be difficult to define the scope of services and therefore adhere to budget and schedule constraints. Clients also often struggle to define their requirements within their budget and schedule constraints with preferential wants and eclectic interested users and stakeholders. The Sydney Opera House is a clear example (EOI, 2012). Also highly creative personnel are driven by altruistic purposes and it can be very difficult for operational managers and project leaders to monitor performance and be certain of the outcome.

Urban design is the blending of urban architecture with public realm, landscaping and urban engineering. The built form is to be considered as urban sculptures and design firms are at their best when there is the integration of the design disciplines. The civil and civic structures should represent their purpose. Akin to the Greek mythology of Atlas holding up the world, these structures should demonstrate their purpose, ingenuity and robustness to the community. In these circumstances, design is more than simply complying with the client's stated project requirements. Astute design firms understand how the proposed development will integrate within the built environment. Their legacy is testament to the design outcomes.

Consequently, many large professional services firms include urban planners, architects and engineers to provide fully integrated design solutions. By example, in the tendering of Terminal 5 development at Changi Airport in Singapore, the client requested a consortium of signature and local architects and engineers to provide a fully integrated design. Operational managers and project leaders need to understand and appreciate the complexities with the diversity of professional disciplines in order to be able to harness the technical qualities as well and the non-technical qualities of the business unit or project team. The complexity of these people and cultural issues are discussed in detail later in the book.

3.3 Projects-oriented firm

This section describes the role of the consultant as a major projects designer. The most common forms of project delivery for these design consultants are:

- Design for the Project Owner
- Design for a Contractor: Design and Construct (D&C) or an Engineer, Procure, Construct (EPC) (Fédération Internationale Des Ingénieurs-Conseils, 2016)
- Design with the Owner and the Contractor: Alliancing (Department of Infrastructure and Regional Government, 2015), Integrated Project Delivery (IPD) (American Institute of Architects, 2017), Early Contractor Involvement (ECI) and other forms of collaborative contracting.

The Owner's decision as to the form of project delivery will have a different design-risk allocation. Designing directly for the Owner may ensure form and functionality, but may not fully exploit innovations and constructability. Also any design deficiencies will be borne by the Owner, or be passed through to the designer.

D&C contracting transfers the design risk to the constructor and, with design being undertaken competitively during the tender phase, can yield innovations and constructability enhancements. However, if the project requirements are not explicitly defined, transferring this risk to a D&C Contractor can mean the Owner does not always get what they thought they were buying. The risks associated with D&C tendering and contracting are discussed in more detail later in the book.

Collaborative contracting aims to blend these two paradigms such that the Owner, designer, constructor and operator work together in an integrated manner to achieve the project objectives. The commercial framework is established to drive common objectives, collaborative behaviours and share in the project success or failure in an equitable manner.

Some of the principles with a collaborative delivery methodology are:

- Trusted collaborative partnership with the Owner, jointly wedded to the project outcomes (win/win)
- Strong safety culture and processes

- Strong relationships and engagement with key stakeholders, such as related government agencies, local council, interest groups, etc.
- Developing works orders and budgets to maximise involvement of the local industries
- Wedded to project outcomes, accountable for delivery
- Demonstrate value-for-money (VfM) through contesting all trade packages with local subcontractors
- Social procurement, maximising employment and engagement with the local community
- Flexibility to respond to emerging project challenges
- Complimentary skill sets with the Owner's project team
- Effective integration with the associated contracts
- A commercial framework that helps form a united project culture and drives the right behaviours to achieve the project objectives, with interim and overall targets
- Preparedness to put 'skin-in-the game' to share in the project successes.

Working for a D&C contractor is not for the faint-hearted designer. Fast-tracked projects are challenging, given they are predominantly programme driven and require strong management as well as oversight and governance. They are best suited when the design is well-defined and locked-down, and easiest when they are essentially document and construct contracts. The more open or performance-based the Owner's Project Requirements (OPR), the more problematic the project is likely to be. The better the scope and expectations are defined up-front, the greater the likelihood of project success. An old adage in fast-tracked construction is 'design as tendered, construct as designed'. Design change detrimentally infects project success. This is discussed in greater detail later in the book.

There will be many challenges to be addressed over the journey of fast-tracked projects. Collaborative contracting provides the opportunity for the designer to be far more influential in the project outcomes than in a D&C contract.

Managing a series of major projects will require a different skill set from a portfolio of small and medium-sized assignments. The cost of managing the project will be higher and so too risk contingencies. Also utilisation and therefore retention of key personnel between projects needs careful consideration. They can be utilised on development of systems and processes, harnessing lessons learned and planning for the next proposal. However, highly experienced personnel are generally not energised during this down-time. So often they are deployed into other jurisdictions and that presents challenges of mobility and transitioning back to the home office. Therefore managing utilisation outside of the project duration is a core challenge of projects-oriented professional services firms.

The latter chapters in this book highlight the differences between corporate governance of a portfolio of projects with project governance of a select number of major projects with abnormal scale and complexity, representing potentially high-value, but also high-risk.

3.4 Projects-services oriented firm

A projects-services firm generally provides project management and technical services to the Owner throughout the project lifecycle. This role is seen as the quintessential trusted partner providing project-delivery expertise allowing the Owner to focus on their core business activities. These project teams are generally sourced with highly experienced personnel and challenges arise due to lack of leverage of the other disciplines, succession planning and knowledge transfer.

Often these personnel are embedded in the Owner's office and that creates the challenge of maintaining the 'umbilical cord' back to the home office. With large projects or programmes of work these personnel are often away from the home office for considerable periods. Although the operational managers may be content with the utilisation of highly paid key personnel and the income stream, again the balance between project and corporate governance is complicated in the sharing of business strategies and initiatives with the remote staff.

Sometimes these project teams suffer from 'Stockholm syndrome', the term applied after a group of hostages sided with their terrorist captors. While a unique project culture should be formed with the Owner, it should not be elitist or show excessive hubris having disregard for the alignment with the organisational culture and business processes. Again this highlights the need to align project governance with corporate governance.

In summary, many professional services firms have struggled to appreciate the balance between these different perspectives and harness the synergies to manage a diverse professional services firm. Using small-practice expertise on major projects is unlikely to lead to success. Conversely, managing a diverse portfolio of small and medium-sized projects requires different knowledge, skills, experience and attributes (KSEA).

4 Marketing and communications

It takes many good deeds to build a good reputation, and only one bad one to lose it.
(Abraham Lincoln)

This chapter describes the marketing functions of a professional services firm. Marketing is defined as understanding the client's needs and wants, aspirations, expectations and requirements. Business development is the process of building client relationships and positioning. Sales or selling is winning the proposal or tender and these elements are discussed in the following chapters.

4.1 Marketing

Marketing is about creating awareness of the firm as the consultant of choice, as identified in the indicative organisational strategy map in Table 2.2 (Levinson, 2004). It's what we do to identify and attract new clients and satisfy and retain existing clients. We offer our clients services through local offices with the strength, depth and diversity of our worldwide practice behind them. We focus on outstanding service delivery. So, in essence our brand is the client experience!

It is not a logo, or a green dot above the firm's name. It is also not a capability statement.

It is about building a brand and reputation based around our people behind our client's success. It is about telling stories. That is, telling stories through technical seminars and conferences, client interviews, any interactions, third-party endorsements and now social media.

Malcolm Gladwell (2000) in his book, *The Tipping Point*, described how little things can make a big difference. He defines a 'tipping point' as the moment of critical mass, the threshold, the boiling point and the mysterious ways that ideas, products and messages and behaviours spread like viruses. In this context, it is the stories told by our 'frontline' of key professionals that will create the brand of the firm. It is therefore recommended that these key personnel are armed with success stories.

Again adapting the 4Ps of marketing (Kotler, 2012) to a professional services firm, the fundamental elements of marketing are as follows:

- People
- Process
- Product
- Price.

With the first P, having the 'frontline' of key personnel is the core requirement. Thus the broader the frontline of key personnel the greater the connectivity to the market.

However, broadening the frontline is a 'chicken and egg' issue. By this we mean recruit the right people and they will win the work. However, this often takes time and the lag time can be an expensive investment, if not fully integrated with the existing personnel in the firm. Organic growth is challenging particularly if starting off a low base. Acquisitions provide the opportunity for established and proven client relationships, technical expertise and pipeline of opportunities. Mergers and acquisitions (M&A) are a key business development opportunity and are discussed later in Chapter 7.

So, the successful professional services firm will seek to broaden the frontline of key personnel and arm them with the stories to sell the brand of the firm. In consulting firms, the frontline are often seller–doers. However, for larger opportunities there needs to be dedicated business development, or external sales personnel and this can be a significant investment.

There are many industry surveys that consider the brand awareness of the firm (Beaton Research and Consulting, 2017). Typical attributes that are surveyed include:

- Technical expertise
- Reliability
- Responsiveness
- Understanding of the client's industry
- Quality of documentation
- Ease of doing business
- Communications
- Price and cost consciousness
- Innovation
- Commerciality of the advice
- Involvement of partners/principals
- Likelihood of repeat business
- Brand awareness.

Clients will value these attributes differently. Value is in the eye of the beholder. Clients buy, they are not sold, so the brand of the firm should be consistent with our value proposition.

Often, professional firms struggle to manage the three dimensional nature of the organisation, being:

- Sales
- Operations
- Technical.

What does the firm want to be known for? Are we best in class for operational excellence, client relationships, or technical excellence? Most firms are primarily known for one of these three elements.

The professional firm that can distinguish between corporate management and client management will provide the 'tipping point' for communicating the brand across the market sectors. Often operational managers of professional services firms will focus on the bottom line; profit and cash. However, these are lag indicators and by focussing solely on the outcomes, this fails to comprehend and appreciate the inputs of people and clients.

By client management, we mean having a true appreciation and engagement with our preferred clients. This means we need a strategic and targeted approach to our client management. Engaging client management is discussed in the next chapter and effective organisational-matrix management is discussed later in the book.

4.2 Communications

Communication is not advertising. Communicating the brand is a combination of telling these stories verbally face-to-face and through marketing materials, presentations at conferences, publications, newsletters, submitting proposals and now through the use of social media.

Capability statements and brochures have a limited purpose and are not a substitute for personally selling the value we can offer our clients.

Social media has provided the opportunity to share these stories to more people and in real time. However, as a people business, these stories will be far more impactful if told through the frontline of key personnel. Again as a people business, the brand is communicated through interactions with our key personnel, as well as through the application of new technologies.

Also, in the delivery phase, the manner in which we communicate with our clients, both verbally, non-verbally and written reinforces the brand. The brand of the firm needs to be consistent through the project life cycle. This is known as honouring the brand promise.

5 Business development

This chapter outlines the process of Business Development (BD), or outside selling. The previous chapter outlined the need for a 'frontline' of key personnel to create the brand as the consultant of choice. This chapter outlines the process for building meaningful client relationships based on trust. The following chapter describes the process for preparing a winning proposal.

5.1 Context

An indicative portfolio of a large professional services firm is outlined in Table 5.1. As discussed in Chapter 3, balancing a consulting or projects firm highlights the diversity of the project portfolio and therefore the differing planning and managing activities required. On one hand, almost one-third of the firm's profit can be earned by a small number of major projects. On the other hand, a portfolio of small projects 'swim through' the firm each year and need to be managed with little burden. The 'swim lanes' for each project type will require a different approach and this approach should be on a risk-based approach, as described later in the book.

This means that the application of organisational management and project leadership needs to be adapted to the scale and complexity of the project, or portfolio of projects. And also the manner in which we pursue these opportunities, either by client or by project.

Many professional firms use the same salary multiplier for selling their costs, irrespective of the different costs and risks of selling and managing. However,

Table 5.1 Indicative portfolio of projects by fee scale

Net fee range (US$)	*% by #*	*% by fee*
<$50k	65	8
$50k to $500k	21	25
$500k to $2m	7	22
$2m to $10m	4	16
>$10m	3	29
Total	100%	100%

each project represents a different scale and complexity so the salary multiplier and therefore profitability should be different, recognising the risks and the client's appreciation of the value we bring.

While we should diligently develop a bottom-up first principles estimate of the cost of delivering the services, we should always consider what the client is willing to pay, top-down. We should always endeavour to sell value (top-down perspective) and not costs (bottom-up perspective). How do we know what the client values? The next sections describe the process for determining this.

5.2 Client relationship management

The process of client relationship management (CRM) is generally not well understood or implemented in professional services firms given the complexity of managing a multi-dimensional business of technical service lines, local, national and international client and different client types. While many key professionals will have close working relationships with their existing clients, the process of client engagement is not well understood as we seek to blend personal relationships with organisation relationships. Also, distinguishing between an acquaintanceship where we know the client, to a relationship where we know what's on their mind.

Table 5.2 provides some of the issues to be considered in order to continually enhance and improve our approach to effective client-relationship management.

Table 5.2 Indicative CRM issues

Issue	*Response*
Strike rate	Often the strike-rate by number is higher than the strike-rate by value. This is due to the fact that smaller commissions are generally sole-sourced, are a trade-up from a previous assignment, or we are in a preferred position. Small proposals should be considered as a portfolio or panel arrangement and reduce the cost of selling.
Client classification	Often professional services firms use terminology such as key, valued, aspirational and project-by-project to differentiate the client's importance. However, this thinking is often based on lag data and may not represent the pipeline of future prospects and proposals. Using the Pareto principle, 20% of our clients are likely to yield 80% of our future revenue, which are the clients for whom we should: • Invest our limited BD dollars on; i.e. aspirational or very important clients • Ensure reliable delivery to be confident of client retention and repeat commissions. Rather than classifying clients, as a mature firm we should ensure our client retention as a core responsibility of the Project Governor and then focus our BD efforts on the targeted very

(Continued)

Table 5.2 (Continued)

Issue	*Response*
	important clients. That is, the CRM effort is significantly reduced and focussed on client maintenance (aka farming) during project delivery where we have established relationships. Hunting new clients should be undertaken strategically, as there is a high cost and often a long gestation.
CRM role and responsibilities	Each client manager will often have a different approach to the role, which could be appropriate for the various clients. However, there should be consistency of approach and the role description should be loaded on the BMS. Who the client manager reports to is often not clear, given the three-dimensional matrix of clients, disciplines and operations. The client manager should be held accountable for the sales target, or revising the sales target on a quarterly basis if the market conditions have changed. The role of the client manager is as the key interface with the client, or the coordinator to manage the client team and the 'frontline' of senior professionals. There is often confusion over the role with national clients and what is a local issue or a national issue. Local client mapping using the traditional relationship model generally works well, but becomes more complex with national and international clients.
Client team	On many occasions, when the client relationship is limited just to the client manager, there is limited engagement with the broader frontline and the limited deeper zippering even down to the graduate and emerging professional level, to seed the relationships for the future. There is often limited discussion about succession planning and developing the client relationships for the future.
Return on Investment (ROI)	Where should we invest our limited BD dollars? In the client development, pursuit, proposal or delivery phase? Working backwards, it is often suggested the project leaders are not undertaking sufficient client-retention activities during delivery phase. This highlights the need for greater project governance, reviews and assurance. In this context it would mean more client-satisfaction interviews, early client-disagreement resolution and project health checks. In the proposal phase, client interactions during the tendering period are often limited. The pursuit phase should provide the best opportunity to gather technical and non-technical client intelligence to be used in the proposal and the building of a trusting and enduring client relationship in the delivery phase. Activities during the client development phase are often ineffective in today's economic climate and probity restrictions.
Client data capture	There is often a lot of criticism of the CRM systems whether they are ineffective or not well utilised. An effective CRM system will vary from minutes of meetings, sharepoint notes, to spreadsheets, and to uploads on the sophisticated sales databases, such as Sales Force.

Issue	*Response*
	With an inclusive culture, there is a need for better sharing of client interactions, intelligence and knowledge in an open, inclusive and transparent manner. It is recommended that there be a consistent system for sharing client intelligence, accessible via the BMS.
Approach	Many client managers recognise the need to 'zipper' the client organisation, mapping decision makers and staff that had close working relationships with key client personnel. However, this mapping exercise is often not formalised and appears to stop at the client manager. That is, the upstream senior management are often not part of the client engagement planning. Although senior executive interaction may not be directly relevant to an immediate pursuit, if the brand is reflected by the senior management, then key clients need to know the senior management well enough to pen a testimonial or address a critical issue.
Relationships	There is often feedback that the project team had good technical knowledge but were devoid of the BD acumen. There is often limited discussion about alignment of personalities (e.g. Graves model) with the client counterparts and this appeared to be an oversight in developing meaningful client relationships. There is often limited discussion about our approach to developing close working relationships. They can appear to be more transactional than enduring and intimate, based on trust and respect. Just because we know the client doesn't mean we understand them. There is a significant difference between having an 'acquaintanceship' and a relationship.
Forecast revenue	The client manager often has the annual report to hand with knowledge of the client's strategic plans for the next three years and the drivers for expenditure on consultants. However, there should be formal quarterly reviews of sales targets. Sales targets are driven by populating a pipeline of known opportunities which is readily accessible for government agencies that publish their capital expenditure plans and other key clients. However, there is often little analysis of the unknown opportunities, except by back-casting revenue from key clients.
Collaboration	Criticism sometimes arises when one business unit provides the capability to the client of another business unit. In other words, if the client was owned by another business unit, then it appeared that client was considered second rate. This is an organisational issue. In this case it is suggested that the client manager be from the capability business unit and the project leader from the client business unit to purposefully drive this cross-fertilisation.
Panel agreements	Many of the target clients now use panel or framework agreements. On many occasions the client managers will consider they are winning more than their fair share.

(Continued)

Table 5.2 (Continued)

Issue	*Response*
	All panels are subject to selective evaluation. The best way to win more work from the panel is outstanding service on current assignments. Just as Abraham Lincoln said, 'it takes many good deeds to build a good reputation, and only one bad one to lose it'. Per the comment above, the role of the project leaders is ensuring client retention.
Differentiated value proposition	For many of the target clients, our competitors are the usual peer group. Our differentiation is not always clear. Price? Technical excellence? Service reliability? Client relationships? All of the above? What is this firm known for? What does our brand represent? Do the project leaders reflect our brand? Have we offered the project leaders with the right KSEA that the client would respect?
Trusted partner	Most professional services firms recognise the concept of developing trust-based relationships with their preferred clients, as outlined in David Maister's book *The Trusted Advisor* (Maister, 2001). However, it is the application of our technical expertise that requires further development of our key personnel to develop these trusting relationships.
Pricing strategies	Selling cost at a set salary multiplier using a bottom-up estimate should be used for benchmarking purposes, along with a top-down value-based assessment. Using the same salary multiplier for all staff in all disciplines or business units does not reflect the various markets. Also, the revenue share arrangement between business units within the work breakdown structure (WBS) needs to reflect accountability for project performance.
Client satisfaction	Often client satisfaction surveys are not undertaken; only when it is likely that we will receive positive feedback. The Net Promoter Score is a useful tool, but often fails through its lack of implementation (Net Promoter, 2017).
Owner not the client	On many occasions our relationship management is with the Owner, but our engagement is via an intermediate party such as D&C contractor, FM contractor, or Project Manager. While we may need to gather intelligence from the Owner to gain a position with a D&C contractor, the way we market to contractor clients is very different from marketing to end Owners, such as government agencies. The CRM engagement should provide for personnel that relate to contractor clients. Contractor clients often take technical capability for granted and engage consultants that are reliable and they can trust to be 'in the kitchen' with them during the challenges of fast-tracked delivery.
Organisational constraints	As a consequence of the GFC, many professional services firms have become organisationally introspective rather than client centric. With many organisational re-structures over recent years there is often lack of clarity with the business development

Issue	*Response*
	activities being a local, client or a national capability responsibility. Also confusion over corporate and project governance and what is a business unit responsibility and corporate responsibility. While many organisations struggle with matrix management, how we are organised is not relevant unless it impedes our service delivery. A client-centric organisation will adapt the organisational structure to meet the client's needs. Also, there needs to be clarity between a centralised corporate function and the responsibility of a business unit. This is akin to 'collaborative federalism' and the USA, Australia or European Union.
NPT balanced with PT	The CRM function often withers when people are busy. Non-project time (NPT) is our investment in the future, to be invested either in our clients or our staff. With a culture of utilisation or billability, there is uncertainty with investing NPT in client management and retention activities. Under-investment will prejudice the longer term. However, expenditure without accountability is a waste. These investments need to be undertaken in a planned and managed manner, with the same operational rigour as the delivery phase. Again, 'what gets measured gets managed'.
Seller–doer model	As explained earlier, many advisory or consulting operations adopt a seller–doer model. However, with technical specialists they often focus on selling their technical expertise rather than responding to the client requirements as specified or their preferences as gleaned prior to receiving the RFP. These proposals need oversight from the client managers to ensure we submit a winning proposal.
Tender review	High-value, high-risk, strategic tenders often do not adequately pass through an appropriate review process. These tenders should pass through three stage gates with strategic review by an independent review panel unbiased by the tender and able to think laterally outside the box. They should also be used in the delivery phase as a shadow evaluation panel in preparation for the interview and Q&A phase. This tender-review process is discussed in more detail in the next chapter.
Corporate approvals	Some client managers complain about an overly onerous corporate approval process and gaining approval to submit a tender, particularly when there are short client timelines for consulting assignments. Although the approval process can be complex, a well-structured proposal manager should plan for the corporate approvals that should be reasonably expeditious if there is a precedent with our preferred clients. The approach to internal corporate approvals is a reflection of how we interact with our clients and other stakeholders. Internal approvers are just one of the stakeholders that need to be managed.

This is an enduring challenge and the successful professional firm will continue to enhance its relationships with its key clients.

5.3 High-value clients

This section discusses some issues to be considered with targeting high-value clients. In this context, high-value is defined in terms of the out-turn salary multiplier or the quantum project gross margin. High-value can be achieved through various sources, such as:

- clients paying a premium for the (perceived or actual) value we provide
- productivity gains where the out-turn costs are less than originally estimated or forecast
- a risk/reward arrangement such as enhanced margin on achieving agreed KPIs and gain/pain sharing with the Project Owner.

While there is no set formula for achieving a high return from key clients, similarly we are not guaranteed to receive another high yield on repeat commissions. Some common themes from high-value clients are as follows, using the previous strategy map format.

Client perspective

We will invest in the client relationship – we understand their industry and market forces. We understand their business and their personal drivers.

We will share and collaborate – we develop a shared understanding with the client. We harness our local, regional and global best practice. We utilise the full depth of the firm's expertise. We engage the client in the process and celebrate successes.

We will put ourselves in our client's shoes – no surprises! We keep the client informed at every stage. We listen to their needs and are flexible and responsive. We honour the promise made at our proposal.

We will be focussed on client care – we explain what is important to them. We share the risks and opportunities with the client. We provide commercial and technical solutions which are outcome focussed. We are innovative and seek to add value.

In essence, their success is our success with a win/win outcome.

Process perspective

We will have a risk appetite and preparedness for risk. The proposal is generally a fixed fee based on detailed scope definition and assumptions, along with programme and quality outcomes. The fee will normally have a defined payment structure.

The variance between the revenue and costs generates (negative) work-in-progress (WIP), which is managed as contingency throughout the project. A risk and opportunity register is monitored regularly, where risks will be mitigated and opportunities exploited. When we manage well, this negative WIP then becomes enhanced margin, or write-up, and everyone profits from making a profit.

People perspective

We will have project leaders who are commercially astute and seek high-yield outcomes. They will be most often experienced with exceptional planning and project management skills, including recovery for scope variations and contingency management.

The project team will be dedicated and focussed on the outcome. No time is wasted on contractual arguments. All the energy is focussed on what the client needs (and/or wants). Everyone on the project team is genuinely interested in pursuing the client's interest.

The project leader will be able to achieve break-through performance, which is far above the traditional business-as-usual, by harnessing the positive energy with a united project team, creating a high-performing team culture.

The project governors will provide appropriate supervision, guidance and oversight of the project to diffuse any disagreements in a timely manner and enhance the services using their KSEA.

So in summary, high-yield clients recognise the value we bring and reward us through enhanced margins and repeat commissions. For an enduring relationship, there can only be win/win outcomes.

5.4 Client development

> Feedback is a gift, provided it is given with the right intentions.
>
> (Tim Ellis)

This section identifies the attributes of our key personnel in developing enduring relationships with our clients.

Client feedback reinforces that client loyalty is derived from the way we work with our clients, as well as what we deliver. We strive for the highest standard of technical excellence. However, client loyalty is derived from technical excellence and outstanding service delivery. This is the essence of our service delivery and culture objectives in the organisational strategy map provided in Table 2.2.

The relationship with the client is at its best when we act together to focus on the client's needs (aka defined wants) and respond with passion so that the client is enjoying an exceptional experience.

We often operate in a very competitive marketplace and it is generally accepted that the relationship with our preferred clients and the provision of outstanding service delivery will clearly be our point of differentiation.

Cultural and behavioural change

We know from market research and from client and staff feedback that adapting our behaviours is the first building block in creating a unique client experience.

Table 5.3 Indicative relationship behaviours

Behaviours	*Client perspective*
We invest in Client relationships	We understand their industry and market forces (PESTEL). We understand their business and their personal drivers.
We share and collaborate	We develop a shared understanding with the Client. We harness our local, regional and global best practice. We utilise the full depth of the firm's expertise. We engage the client in the process and celebrate successes.
We put ourselves in our Client's shoes	No surprises! We keep the Client informed at every stage. We listen to their needs and are flexible and responsive. We honour the promise.
We are focussed on care value	We explain what is important. We share the risks and opportunities with the client. We provide commercial and technical solutions that are outcome focussed. We are innovative and seek to add value.

This change in mindset is part of the high-performance strategy and is underpinned by our values. Some would regard this as our brand. Table 5.3 outlines some of the desired behaviours in developing enduring relationships with our preferred clients.

In his book *The Trusted Advisor*, David Maister (2001) suggests that getting hired is about earning and deserving trust. What would make you trust someone?

To earn a relationship, you must go first – give a favour to get a favour. The one you are trying to influence must visibly perceive that you are willing to be the first to make an investment in the relationship. To make anyone believe something about you, you must demonstrate – not assert – illustrate – don't tell. Trust-based relationships are based on robust dialogue, ownership and accountability and a focus on results that define the essence of the relationship. Adapting the trust equation is:

$$T = C \times R \times Q / SI$$

Trust is earned by credibility (C), reliability (R) and the quality of the relationship (Q), which can be eroded by self-interest (SI). Given that credibility and reliability are often business-as-usual functions, the quality of the relationship with the client can be the defining element. That is, the quality of our deliverables (technically) as well as the quality of our relationships (non-technically). What does non-technically mean?

Clients hope for a relationship that is open and transparent, to remove any barriers – no hidden agendas. No backwards and forwards, lobbing documents over the fence, just one united team with common objectives.

No time would be wasted on contractual arguments. All the energy would be focussed on what the client needs (aka defined wants). Everyone on the project team is genuinely interested in pursuing the client's interest.

This provides a culture of constantly seeking new and better ways of doing things while existing systems, processes and procedures are being enhanced and continuously improved.

We know and can demonstrate that the client is receiving excellent value in the relationship. This methodology (PwC, 2017) is summarised in Table 5.4.

Some indicative steps are suggested as follows:

- Communication – talk early, talk little, talk often
- Kick-off workshop – hold a kick-off workshop to develop what success would look like at the end of the project and celebrate appointment and lessons learned on success
- Celebrating the client experience – having completed the report on Friday night, after extensive after-hours work, hand-deliver the report and suggest drinks and nibbles to celebrate
- Client Engagement – host the client meetings in the office and involve the client in the process and with all project team members
- Reward and recognition – ask the client to attend the staff meeting to present the client experience award
- Reporting – use the monthly invoice and a brief progress report and seek immediate feedback of performance. Remember, feedback is a gift.

Table 5.4 Indicative client engagement methodology

Methodology	*We invest in client relationships*	*We put ourselves in our clients' shoes*	*We share and collaborate*	*We focus on client value*
Prepare	Take a long-term view on relationships with (preferred) clients	Understand the clients' business and personal drivers	Share insights and expertise with clients	Seek unexpected insights for the client
Engage	Spend time with the client and build relationships	Validate expectations with the client	Act as a united team with the client	Commit to service levels with the client
Deliver	Focus on the service experience and technical delivery	Make outputs client-focussed	Celebrate service excellence	Establish and demonstrate value
Enhance	Make our networks work for the client	Seek independent feedback	Take time out to consider service improvements	Review performance with the client

Game-changing transformation

Lessons learned for recent relationship-based or collaborative-contracting (e.g. alliance contracting) experiences are that performance is far above the traditional business-as-usual by harnessing the positive energy with a united project team, creating a high-performing team spirit.

The principal challenge for the project team is to develop a united team culture that will break through traditional barriers and paradigms achieving game-changing outcomes (Alchimie and DLA Piper, 2003). This culture should yield solutions that are regarded as possible, but the team doesn't initially know how to achieve them. It will require a new way of thinking and the project team needs to be committed to making it happen.

Therein, the need for significant mindset change to develop a high-performance leadership, management and wider project team that are all 'on-board the bus'. Breakthrough performance is less about individual team members and more about galvanising people around clear, seemingly improbable challenges.

The performance spectrum can be described as follows (Table 5.5).

Relationships for the twenty-first century

It is generally recognised that future operational managers and project leaders will operate in environments quite different from those they first entered. Successful project leaders will master a range of hard and soft skills. The business-as-usual hard skills include scope, time, cost, quality and contract management. The softer skills including managing the client's expectations, complex interfaces with stakeholders, communication and people management will be more prevalent than in the past.

Table 5.5 Indicative performance spectrum

Spectrum	*Description*
Outstanding	• Breakthrough performance • Regarded as possible, but don't know how to achieve it • Will require new way of thinking • Committed to make it happen
Stretch	• Know how to achieve it • Incremental change • Would require systemic and repeated success
Business-as-usual	• Owner's minimum acceptable expectation of performance • Level of performance is commensurate with best-in-class performance
Poor	• Below business-as-usual performance
Failure	• Significant impact on project • Would require systematic and repeated failure

Some of the issues to be considered that will impact on the client relationship are:

- Three generations interacting
- Longer working life
- Workplace flexibility
- Personal issues
- Diversity of gender, age, cultures and KSEA
- Inclusiveness of diverse stakeholders and perspectives
- War for talent and resource constraints
- Under-servicing in resource-constrained markets
- More demanding and diverse stakeholders
- Increased regulation and approval processes.

The psychology of the relationship

Much work has been undertaken on understanding individual preference and personality types with psychometric assessments such as Myers Briggs Type Indicators (MBTI) (The Myers Briggs Foundation, 2017), Enneagrams (The Enneagram Institute, 2017), Human Synergistics (Human Synergistics, 2010) and the like.

Dr Clare Graves (2017) described the psychological relationship by colours and this is relevant to the interaction of professional services firms with their clients.

- BEIGE – natural, instinctive
- PURPLE – traditional, conventional
- RED – self-asserting, dominant, power, egocentric
- BLUE – conforming, absolute, obedient to higher authority
- ORANGE – pragmatic, achiever
- GREEN – consensus, collaborative, relativistic, situational
- YELLOW – independent, niche
- TURQUOISE – experimental, innovative.

Astute project leaders will develop more adaptive personalities to achieve better alignment with their clients and partners. In some instances, appointment of the project leader may be on a relationship basis rather than technical competency. This will obviously require a team approach with a supportive project governor, project controls and strong technical leads.

Many government agencies are green style. The proposal submission often requires the completion of numerous returnable schedules and often has a large evaluation panel with a very structured assessment regime. Conversely, many private clients are red style. The proposal can be a series of short slides outlining what success would look like and then the commercials negotiated in a collaborative manner in good faith.

This assessment of the client's preferences can also be useful in structuring the tender submission. Red clients will want a short and concise proposal, perhaps a few powerpoint slides. Green clients will expect the extensive number of returnable schedules to be completed in minute detail.

The emotionally intelligent leader

Many of our projects are extremely challenging and make constant demands upon our mental and emotional reserves, sometimes to the point of depletion. This happens in both our personal and professional lives. To be able to survive the eroding and pulverising effects, we need to develop mental toughness and emotional resilience. The core definition of mental toughness (there are a multitude of definitions) is the ability to remain focussed, resolved and confident even when the challenges of life are trying to beat you down. Emotional resilience is defined as the ability to spring back from such challenges by being able to remain cool, flexible and positive, no matter what is going on in your life. Without these qualities, we will suffer fatigue, decreasing well-being, decreasing effectiveness and, of course, poor client relationships. Resilience and persistence are one of our core values as identified above in our beliefs.

Technical skills alone are no longer sufficient. Knowledge, skills, experience and attributes (KSEA) are the foundations for our project leaders. Project leaders need to have a strong emotional component, to have high levels of self-awareness, maturity and self-control. To be able to withstand the heat, handle setbacks and when the lucky moments arise, enjoy success with equal parts of joy and humility. There is no doubt that their emotional qualities (EQ) are equally, if not more important than intellectual or technical competency.

Some of the EQ attributes (Newman, 2009) of high-performing project leaders are:

- Independence – ability to earn trust with single-point accountability
- Assertiveness – provide clear and concise decision making
- Optimism – positive outlook
- Self-awareness – aware of the influence on others including the client and stakeholders
- Self-confidence – high levels of self-regard and self-esteem
- Interpersonal relationships – strong working relationships with the client and other team members
- Empathy – can stand in the client's shoes and understand their perspective
- Energy – creates a passion and energy that inspires the client and the project team.

It is regarded that the utopia of true human potential of project leaders is achieved by harnessing their intellectual capability (IQ), with their technical expertise (TQ) and enhancing the combined result through emotional intelligence (EQ) and the spiritual, values, principles or beliefs qualities (SQ). The human potential equation is:

Human potential = (IQ+TQ) x (EQ+SQ)

Table 5.6 Suggested steps for instilling cultural change

ID	*Activity*
1.	Undertake research on the client experience and other service delivery dimensions
2.	Hold focus groups to develop ideas as to what represents an exceptional client experience
3.	Nominate cultural change agents within business units
4.	Capture war stories and testimonials of outstanding service delivery
5.	Re-examine and develop client-feedback mechanisms
6.	Develop promotional material to raise the awareness in the office
7.	Develop reward and recognition strategy
8.	Consider psychological training modules for selected project leaders
9.	Incorporate relationship metrics in the key client plans
10.	Incorporate the client-experience drivers in the project-management accreditation process
11.	Inculcate a high-performance environment within the business units
12.	Develop promotional material to be used in business development and proposal documents

There is one absolute certainty in this dynamic: when project leaders develop mental toughness and emotional resilience their leadership skills are enhanced and they become an energised magnet for other people (clients, partners and the project team) who want to draw strength from those character traits and human qualities. This is called charisma!

Presentation, personality and promotion

This chapter has explained that today's consultant is not only focussed on what we do, but how we do it. Our deliverables are a combination of written, verbal, auditory and callisthenic. So for every interaction with our clients we need to exude these EQ attributes. Often clients appreciate only a small portion of our content. It is analogous to a good joke. It's not just the content, but also the presentation. Non-verbal communication can often be more important than what is being said.

Key personnel that exude this confidence and charisma will have far better engagement with their clients than the technical expert that espouses to be the smartest person in the room. How we give our advice can be far more impactful that what we are saying. Again, if a trust-based relationship has been earned, our advice is far more likely to resonate with our clients and partners.

Instilling cultural change for enhanced client engagement

Table 5.6 provides some suggested actions for instilling cultural change for better client engagement.

6 Winning proposals

> The bitterness of poor quality remains long after the sweetness of low price is forgotten.
>
> (Benjamin Franklin)

The foregoing chapters have described the marketing and BD functions of a professional services firm. This chapter describes the process for developing and submitting a winning proposal and the likely negotiation process leading up to contract award. The term proposal is used purposefully rather than tender or quote, to highlight that the submission is an 'offer' to form a relationship with the client and meet the client's functionality, schedule and budget requirements.

An 'offer' means we understand what's on the client's mind, not necessarily just what's requested in the tender documents. The foregoing chapters endeavoured to establish what's on the client's mind before we commence preparation of the tender submission.

A structured tendering process, such as in the public sector, will normally commence through an Expression of Interest (EOI) phase, or selection from a panel or framework, then a shortlisted number of tenderers will be invited to submit a Request for Proposal (RFP). In a less structured process, such as in the private sector, the proposal will be more a quotation, with price being a key determiner.

In both cases, the relationships need to be formed before the RFP is issued, as explained in the previous chapter.

6.1 Tendering to owners

When responding to a RFP, the proposal preparation proceeds through a structured process, outlined as follows through three key stages:

- Stage 1 – Opportunity tracking
- Stage 2 – Proposal development and preparation
- Stage 3 – Client evaluation.

The steps within each stage are summarised in Table 6.1. There are a multitude of other intermediate steps, but these are the main gates, decision, or corporate approval points.

Table 6.1 Indicative proposal process

	Stage 1	*Stage 2*	*Stage 3*
Step	*Opportunity*	*Proposal*	*Client decision*
Aka	*(Suspects, Prospects, Pursuits)*	*(Offer, Tender, Quotation)*	
1	Identification	Review RFP	Evaluation (Q&A, interview)
2	Gathering	GNG to tender?	Negotiation
3	GNG to pursue?	Prepare proposal	Contract execution
4	Pursue	Review and approval to submit	Debrief
5	Prepare for tender	Submit	

The earlier the proposal team identifies the opportunity and undertakes a strategic approach, the higher the likelihood of success. There are generally three types of proposal:

- Compliant, based on interpretation of the tender requirements and assumptions made
- Limited scope, where the scope is aligned to the client's budget
- Added Value, where the client may be prepared to pay a fee-premium based on the enhanced value we provide.

Tendering to government agencies and large private clients is often very onerous with multiple returnable schedules. However, it is contended that the purpose of the RFP is to select the preferred tenderer, not necessarily to perform the services. Often the outturn scope is considerably different from the tendered methodology. Therefore, the RFP submission should focus on both the technical requirements as well as the relationship elements. There is no doubt that as clients are human beings, they will consciously or unconsciously select the firm that they think provides the best value.

For large multi-disciplinary studies and designs, developing a bottom-up fee estimate will often 'over-cook' the fee, given the inherent contingency in each discipline's estimate. All estimates will require a risk contingency as an additional cost. A top-down estimate should be undertaken using benchmark data to determine a competitive fee and then align the definition of the scope and the methodology. In New Zealand and other countries, often the RFP is split into two with non-price submission first and the price submission later, that may or may not inform the evaluation, or inform a value-for-money (VfM) selection.

The evaluation criteria for most RFPs can be summarised again using the 4Ps of marketing (Kotler, 2012). That is:

- Product – past performance, track record, experience, relevance
- People – key personnel, organisational structure
- Process – approach and methodology
- Price – hours and rates.

Sometimes the weighting of these criteria is declared, although more often clients will keep this confidential so they may adjust the weightings depending on the quality of the tender submissions. Clients are significantly better informed upon receipt of the competitive tender submissions, reflecting the market conditions.

With many proposals, 'less is more' and it takes more effort to be more concise. A quote from an unknown source is 'If I had more time I would have written a shorter letter'. A useful technique to ensure the text is impactful is to use the FIBRE acronym:

- Feature – what is unique about this offer?
- Impact – why will this impact the project?
- Benefits – what will be the benefits in terms of time, cost or functionality?
- Relevance – how is the example relevant to this project?
- Evidence – who will attest to this assertion though testimonials, awards or other reference?

Another perspective is that the purpose of the tender submission is to be invited to an interview with the client and the evaluation panel. In these circumstances, the third stage of the proposal process is often when the tender is won or lost. There is no substitute for rehearsing and coaching the team in preparation for the interview and Q&A.

Proposal review and approval

Developing a winning proposal can be very intense with limited time frames and often a bid or proposal team can't see the 'woods for the trees'. A common approach for larger tenders is to utilise an independent shadow evaluation panel for timely strategic reviews and role play the client in the mock rehearsals.

This approach forms the basis of Independent Assurance (IA) across the project continuum, particularly for high-value, high-risk (HVHR) tenders, with the principles of ISO 31000 (International Standards Organisation, 2016). Table 6.2 outlines an indicative proposal review process with staged review gates.

6.2 Tendering with D&C or EPC contractors

The purpose of this section is to provide guidance on some of the considerations for tendering with a D&C Contractor and with allied design partners. It is not intended to be an exhaustive list of considerations, but rather to assist bid teams in the teaming discussions and agreeing the key principles before negotiating services agreements. Most D&C Contractors now have template agreements. This section discusses the application of these agreements, which should be subsequently addressed in our Tender Services Management Plan (TSMP) and Design Management Plan (DMP).

It is proposed to form an exclusive teaming arrangement with the D&C Contractor for the project. The following provides some of the key principles for these

Table 6.2 Indicative proposal review process

Process	*Description*	*IA function*
Bronze	This first stage is when the most **strategic** decisions are made about win themes, teaming, key personnel, governance and our differentiated value proposition. Arguably this is the most critical review point as it could set us down a pathway of investing in a very good losing tender.	The IA team would act as a challenge team or shadow evaluation panel challenging and confirming the essence of the bid strategy.
Silver	The second stage is about understanding scope, methodology and contract conditions and how they impact the risk contingencies. This assessment should determine to what extent we bid-back to the client with departures or clarifications or assess **risk contingencies**.	The IA team could support the bid team in complex contract negotiations and how attendant risks should be priced or mitigated or managed in the delivery phase.
Gold	The third stage is about finessing the competitive fee estimate and soliciting final corporate approvals.	The IA team would undertake the usual review of the fee estimate using reliable benchmark data.
Inception	The inception review is about assurance of setting-up the project for **success**, as expressed in the management plans. In simple terms, the right people, doing the right things, at the right time.	Plan twice, cut once. Often with a procrastinated evaluation period and contract execution, the gun is fired and the project team commences in earnest, with great haste and due expedition. The first 20% is the most critical period.
Key milestone(s)	The next review(s) is again more **strategic** than a QA audit and can be highly influential in setting the pathway for success.	HVHR projects warrant strategic project reviews at key milestones. There should be continuity between the IA during the tender and the PRP during the delivery phase. It also provides accountability for the review team as they has partial ownership of the tender and dispel the 'seagull' perception on the PRP.

teaming arrangements in order to support the D&C Contractor in submitting a compelling Expression of Interest (EOI) document, be shortlisted to respond to a Request for Proposal (RFP) and, on success, delivery of the project.

In progressing the teaming arrangements for all phases of the project, reference should be made to recent projects with the D&C Contractor. We view the relationship with the D&C Contractor as being a trusted design partner, not a design subcontractor. We have a vested interest in the success of the tender and a long-term relationship with the D&C Contractor based on a portfolio of successful tenders and projects.

We commit to all negotiations being undertaken in good faith. These understandings will be captured in the respective agreements or the accompanying management plans.

EOI submission

In today's market environment the EOI phase may require a detailed qualitative technical assessment, more than simply providing capability statements and marketing material requiring a technical response grounded in the project to address the critical issues and success factors. Accordingly, it is proposed that technical services required in the EOI phase to inform the submission should be reimbursed as pre-RFT services, either under a Purchase Order, or under the TSA in a substantive form, to then be applied retrospectively upon execution.

As usual, there would be no reimbursement for management or business development staff in the preparation of the EOI submission. However, we may consider no reimbursement of these technical services should the D&C Contractor not be shortlisted.

Design joint venture

If it is considered that the project requires two lead design consultants, it is proposed that we and our design partner form an unincorporated Design Joint Venture (DJV) for both the tender phase and, upon success, the delivery phase. This arrangement provides for best-for-project decision making, resource planning and clear liabilities between all parties for all phases of the project.

Commercials

It is expected that the project will be procured as a D&C contract with a lump sum based on a standard form of agreement with the Owner. A portion of the bid costs for an unsuccessful short-listed tenderer may be reimbursed by the Owner in return for any intellectual property rights. We would permit assignment of the intellectual property rights based on appropriate reimbursement of our services.

It is proposed that the firm provide technical services during both the Pre-RFT and RFT phases at discounted hourly rates for all design effort expended, plus project expenses, with the fee forgone as a success fee to be invoiced on the D&C Contractor achieving preferred tenderer status.

The extent of discounting would relate to the scale of the project, the number of tenderers and whether the D&C Contractor is being reimbursed part of their external costs by the Owner.

Other consultants

It is likely that the D&C Contractor will engage Other Consultants. The firm would undertake the role of the 'lead consultant' with responsibility for integrating with these Other Consultants. However, the overall design-management function would be undertaken by the D&C Contractor. We would propose that a detailed scope of works and RACI matrix be prepared by the D&C Contractor to provide clear delineation of responsibilities with the D&C Contractor, the firm and the Other Consultants. Alternatively, if the firm engages Nominated (specialist) Sub-consultants, we should be entitled to the cost of managing, margin and sub-limit of liability.

Scope and budget

We expect the Tender Services to be a combination of Tender Advice in a narrative form, drawings and technical specifications and a risk register. We will work with the D&C Contractor's team to establish a Tender Budget that balances their bid budget with developing a compelling design solution; responds to the RFT requirements; is suitable for reliable estimating purposes; and upon success can be readily developed into a detailed design for expeditious approval and construction.

We would expect the Owner's reference design to be reasonably well-developed, say a 40 per cent design. We will develop a detailed organisation chart and resource plan to estimate a Tender Budget. Based on past experiences and the indicative D&C cost, we would expect the Tender Budget for our scope of services to be in the range of 0.25 per cent to 0.5 per cent of the estimated D&C cost at the discounted rates. There needs to be alignment of limited Tender Advice with subsequent liabilities.

The Tender Budget will be monitored on a weekly basis and any Budget Adjustments will be pre-agreed. It is recognised that tender-phase effort is largely driven by the robustness of the reference design, the requirements of the estimating and construction teams and the Owner's submission requirements. The initial estimate of the Tender Budget will be based on certain stated assumptions. Adjustments to the Tender Budget should not be unreasonably withheld. Similarly, pre-approval for overtime should not be unreasonably withheld.

In addition to the Tender Budget, it is expected that we will be required to support any Q&A during the evaluation phase and preferred tenderer status.

Schedule

It is understood that a pre-qualification will be sought with an EOI submission and then an intensive RFT period, likely to be around 12 to 18 weeks, with a number of interactive workshops with the Owner and their advisors and then submission. As explained above, we expect that to collaboratively provide a compelling EOI submission, to be shortlisted to one of two tenderers, the firm should undertake selected preliminary Tender Services to respond to critical technical issues and challenges. We propose that around 5 per cent to 10 per cent of the Tender Budget be assigned as pre-RFT services to support the technical elements of EOI submission.

We would expect the first phase of a compressed tender period to consider innovations to the reference design and value engineering (VE). After freezing the preferred design solution, the second phase is to prepare well-considered Tender Advice Notices (TAN) for reliable estimating and construction planning purposes. The third phase to prepare tender-submission documentation.

Project office

We recognise that collocation within a dedicated project office provides the most effective collaboration, provided the design and construction teams have suitable facilities and space to allow for collaboration and without disruption to producing timely tender advice. If a dedicated project office is to be secured, it is requested that appropriate IT and HSE requirements are provided in advance of the design team mobilising. We would request these requirements are discussed and agreed at least four weeks in advance of the tender office being established to allow adequate procurement time.

Tender services agreement

We acknowledge that the Tender Services deliverables will be subsumed within the delivery-phase design documentation; however, the liabilities under the subsequent agreement should not be applied retrospectively. Therefore, we expect the insurances and liabilities under the Tender Services Agreement (TSA) to be limited to amounts commensurate with the Tender Services fee.

Design services agreement

We recognise that our delivery-phase fee is not competitively bid and therefore we need to provide a reasonable degree of transparency for benchmarking purposes, on a comparable basis. As a key investor in the tender, we will commit to developing a highly competitive lump-sum fee based on the following key principles. We do not want to inflate our risk contingency with unwarranted risks from the head contract or the Design Services Agreement (DSA):

- We will estimate the lump sum fee, based on alignment of scope with the Tender Services; the estimated number of deliverables; and the design programme to estimate the cost of design management, design documentation, verification, expenses, escalation and risk contingency.
- The lump sum will be amortised over the estimated optimal number of design packages for progressive progress assessment and payment.

- The design packages will be submitted for internal review and then externally in three stages, being preliminary, detailed and for construction with the level of detailing to an industry standard (per the projects referenced above).
- The contractual milestones will be limited to a discrete number of critical design packages and we will be entitled to an extension of time for delays caused by late design inputs and third-party reviewers.
- The standard of care will be to an industry standard (not a broad fitness-for-purpose obligation), consistent with our corporate or project insurance requirements.
- The pass-through of head contract obligations should relate to our design-related obligations and not construction-related risks.
- If the project is of high-risk scale and/or complexity we would support the D&C Contractor in procuring a project-specific PI insurance policy to respond to the head contract obligations, with the DJV and the Other Consultants named as insured parties. We would make a risk-adjusted deduction to our lump-sum fee to contribute to the procurement of this policy.
- We will be able to rely upon design inputs provided by the D&C Contractor including Other Consultants.
- Our liability will be limited to the proceeds actually recovered from our insurances, or a monetary sum commensurate with the uninsured risks.
- We will not change the tendered design unless directed to in writing. Any design change, value engineering exercises or justification of quantity growth and re-design would be additional to the lump sum estimate.
- We will not accept penalties for continuity of key personnel provided there is reasonable succession planning.
- We will not accept exclusions to the liability cap, beyond standard industry carve-outs.
- We will not accept onerous set-off, or payment terms, unreasonable time bars, indirect or consequential losses.
- We establish a governance committee comprising authorised representatives from the D&C Contractor and the firm to govern the performance of the services, the relationship, and for resolution of escalated issues.

Construction phase services

We will provide an estimate of the likely CPS Budget based on past experiences, to be provided at 100 per cent of the escalated standard hourly rates for all design effort expended, plus project expenses, to cover:

- requests for information, design changes, technical queries;
- updating IFC design documentation;
- review of shop drawings and vendor data; and
- provision of as-built drawings.

We may consider discounted rates for site-based personnel recognising lower overheads and/or discounted rates for design staff to defray allegations of design deficiencies in the IFC design documentation, balancing the contingency allowance in our lump-sum fee.

7 Growth through mergers and acquisitions

Profitable growth should be a strategic aspiration of all professional services firms, being a combination of organic growth through improved market success, major projects underpinning the profitability and undertaking strategic mergers and acquisitions (M&A). Many recently listed firms have embarked on an aggressive M&A strategy, supported by their new investors. In these cases, the firms have essentially become a federation of acquired businesses, with little harmony across the sum of the parts. It is recommended that the strategic growth of the firm needs to be a sustainable balance of organic growth and M&A. Even with the best-planned strategies, M&A are often opportunistic and the firm should be alert to these opportunities in its BD activities.

7.1 The M&A imperative

The section considers the M&A of infill specialist business. Scale M&A require a different approach and it is recommended that a specialist advisor be appointed. As the firm aspires to achieve its strategic growth objective, acquisition of infill businesses are an integral component, along with organic growth and strategic recruitment of niche business stream leaders. The aim of this section is to discuss the issues to be addressed in successfully completing infill acquisitions and lessons learned from recent acquisitions.

Infill or tuck-in acquisitions are generally characterised by niche partnerships, owned and managed by one or more partners. The success of the business has generally been through the partner's personal reputation, experience and technical competency. In the development of these skills, in a closed environment, they are generally not well equipped in the art of people management or business processes.

The reason for selling varies, but generally the partners are seeking to realise the value in their business sometime prior to retirement, relieve the compliance and administration burden and focus on key technical matters. Therefore, the integration should focus on new career plans for the partners, not managing the business in a similar form and harnessing the business synergies.

Business synergy

The fundamental reason for acquiring another business is that the combined form should be greater than the sum of the parts; i.e. 1+1 >2. Qualitative business synergy can be considered in terms of:

- Market sectors – diversifying into new market sectors
- Geography – diversifying into new geographic areas
- Vertical integration – providing higher value services to existing and new clients
- Client base – broadening the client base
- Leverage – pulling through existing firm-services offerings (Trojan horse concept)
- Technical expertise – enhancing technical expertise and capability
- Competitor analysis – providing a more competitive response.

The acquirer should be responsible for capturing the value from the business synergies. It is unrealistic to expect the partners to lever the business into the acquirer, develop the synergies and capture the commensurate value through increased revenue and profit. If they could have, they would have in their existing form.

The qualitative business synergies that are often overlooked include:

- Enhanced brand profile through promoting technical gravitas and personal reputation
- Enhanced thought leadership and innovative ideas
- Enhanced professional culture through leading-edge technical expertise; i.e. a business based on leading professionals, not a large multi-disciplinary team of generalists/managers
- Mentoring younger staff and rising stars.

Risks and opportunities

Given the relative scale, the acquirer needs to be prepared to be responsible for the risks and therefore seize the commensurate opportunities. The obvious key opportunities are:

- Increased revenue and profit leading to escalating earnings per share.

However, the indirect benefits often missed or not exploited are:

- Additional management talent
- Cultural enhancement
- Reverse integrating the acquiree tapping into better practices in client-relationship marketing, staff welfare and the like.

Key issues to be considered in the acquisition are:

- The age and tenure of partners
- Competency of second tier of management
- Stage in service life cycle and sustainability of the revenue stream
- Nature of the acquisition – asset or share capital purchase
- Lower overheads and charge-rates of the acquiree and higher relative profitability
- Comparable remuneration recognising tax benefits to partners
- Residual liabilities
- Stability and sustainability of the goodwill and possible impairment using IFRS.

Non-competitive acquisitions

In established competitive market places, it is generally a seller's market. Key listed companies and large privately owned firms are seeking to secure greater market share through M&A. For tax purposes, the partners often pay themselves a nominal salary and distribute dividends to family members, trust accounts or other vehicles. Comparable compensation is often difficult to demonstrate.

The privately owned firms generally merge the two entities together on an agreed equitable equity-swap basis. Generally the mergee receives a significant future capital gain of the shareholding in the larger firm. Owning equity in the larger firm inherently breeds greater buy-in and provides parity with the other senior professionals. Sometimes for better equity, the transaction comprises equity in the acquirer and a cash equalisation payment.

Often acquisitions lead to discontent with the incumbent organically grown senior professionals, which limits integration and maximising the business synergies.

Competitive acquisition

In a seller's market, many partners are endeavouring to cash-in their equity. They realise their successors are unlikely to raise sufficient funds, unless they were to engage the financial support of a venture capitalist.

For listed firms, the most common acquisition model is a purchase sum comprising NTA plus a multiple of EBIT. For many consulting firms, the multiple is generally in the order of four to six; however, much larger multiples have been used when a contractor or facilities management business has acquired consultants to pull-through other parts of the organisation.

Many sellers are appointing business advisers who suggest that the P/E multiple should be approaching that of the acquirer. These processes are effectively run as auctions. A common model is 80 per cent of the purchase sum being provided up-front as a mix of cash and shares with 10 per cent after year one and 10 per cent after year two without any links to the future business performance.

In today's competitive environment, the thirst for senior professional talent and the rising age of partners, it is likely that many infill acquisitions will be competitively tendered.

Management control

Although the acquirer would like to minimise the up-front and fixed payments, with an earn-out based on future performance, most infill acquirees are concerned about the loss of management control. Also they perceive they are taking too much of the risk on achieving the business synergies. In other words, if they could have achieved significant growth in their targeted business streams they would have and reaped the benefits. Often the partner(s) is looking to relinquish day-to-day administration and compliance matters and focus on technical expertise and client relationships. For these reasons the partner(s) should not be incentivised to manage the ongoing business stream.

Integration

To maximise the goodwill, successful integration is imperative. The integration plan starts from day one, when the acquiree is welcomed to the acquirer's family. It is to be recognised that the acquiree goes through an emotional experience, and transitions through three phases:

- Phase One – Acquiree now part of aquirer
- Phase Two – The acquirer incorporating the acquiree
- Phase Three – One firm.

A dedicated Integration Manager needs to be appointed to manage the acquiree through these three phases. As an astute and experienced professional they should utilise an adaptive management style through each transition phase. The Integration Manager should be incentivised on a balanced scorecard basis.

Culture

Given the goodwill represents the ability of the acquiree to effectively deliver the transferred backlog, secure submitted proposals and position for new opportunities, it is imperative to win the 'hearts and minds' of all the acquired staff. This would imply that other senior professional staff of the acquiree should share in the purchase sum and/or the business synergies going forward both in quantitative and qualitative terms.

Most infill businesses are run as partnerships. It is recognised that the behaviours used to manage a small partnership are somewhat different from managing a broad-based multi-disciplinary international business. The transition is a complex emotional process with issues such as ego, status, loss of control and general change. Issues such as compliance, risk management and corporate governance are new concepts for the partners. Induction of these business processes should not overly dominate the fundamental consulting principles of client relationships, people management and technical quality.

It is to be recognised that most infill businesses have a unique and special culture of which they are very proud. The acquirer needs to recognise that each business stream will have its own unique sub-culture. The 'soul' of the acquiree should be maintained. Further to this, the acquiree should be encouraged to enhance the culture of the broader acquirer's business.

Lessons learned from converted acquisitions

Some of the lessons learned from converted acquisitions are:

- No Integration Manager appointed nor an integration plan prepared and approved
- Too much focus on the transaction and not enough on winning the 'hearts and minds' of the future staff
- The due diligence process has been overly onerous for the scale of the business and the potential risks
- Given the lack of an integration, the commercial team have overly dominated the integration focussing on process rather than clients and staff
- Although the partner(s) is incentivised to grow the profit of the acquired business stream, there has been little integration with the rest of the business
- The incentives have not been geared to pull-through other business streams and maximise the leverage
- The partner(s) has been incentivised to manage a larger combined business without appropriate skill or acumen
- The partner(s) has been appointed to a senior management position as a line manager in lieu of client-facing positions
- Lack of attention on the key technical leaders from the acquiree
- Lack of support and buy-in from other business streams given the disparity in the acquisition.

Lessons learned from unconverted acquisitions

Some of the lessons learned for unsuccessful acquisitions are:

- As a listed entity acquiring rather than merging small-scale infill businesses, the firm must recognise the current market environment and be prepared to respond to a competitive process, generally conducted by inexperienced sellers, akin to an auction
- With the rising age of the partner(s), sellers are looking for the firm to take on the business going forward and provide a more statesman-like role for the partner(s)

- The firm should be prepared to take on the challenge (aka risk) of managing the business going forward and realising the business synergies
- The purchase sum should be derived based on historical performance; e.g. NTA plus a multiple of EBIT
- Forecasting for future income and profit are problematic
- EPS modelling is often not adequate when determining a 'full' purchase price and internal rate of return (IRR) and discounted cash flow (DCF) and other models should be employed
- The focus on process is overly dominating the winning of the 'hearts and minds' and harnessing the business synergies.

Suggestions for improvement

If acquisitions are a key platform to achieving the firm's strategic growth objective, along with organic growth, it is suggested that an acquisition workshop be held to determine the process for successfully acquiring infill businesses. Some suggested recommendations are:

- Empower key frontline personnel to identify and nurture acquisition targets and establish the business synergies
- Appoint a M&A specialist team to address the marketing, HR and commercial perspectives to work with the local acquisition manager in successfully completing the transaction
- Progress the acquisition as a project appointing an Acquisition Manager to later become the Integration Manager with an approved integration plan with defined outcomes
- Manage the plan as the integration transitions through the three phases
- Focus on winning the 'hearts and minds' of the acquiree
- Be careful with the vernacular and adoption of the firm's systems and parallel the commercial systems for as long as required
- Respect the 'soul' of the acquired organisation as a sub-culture within the overarching firm's corporate culture
- Capture the knowledge and experience of the existing staff who have been involved in outsourcing, mergers and acquisitions
- Promote the acquiree as a source of talent and other business initiatives.

7.2 The M&A process

Some listed professional services firms have an insatiable appetite for M&A, as this drives their growth and therefore market value and attractiveness from investors. However, for smaller firms a M&A is undertaken when the opportunity arises and the process is often not well-understood. The indicative M&A process is outlined in Table 7.1

Table 7.1 Indicative M&A process

Phase	*Activity*
1	**Strategy** Undertake market analysis to identify potential acquisition opportunities that will align with our strategic plan as recorded in the attached register.
2	**Courtship** Agree a mutual non-disclosure agreement and hold initial meetings to explore alignment of values, principles and culture and the likely business synergies in a combined form. Look at exploring joint opportunities to work together. Appoint an Acquisition Manager.
3	**Synergy study** Hold workshop(s) to establish business synergy in a combined form; i.e. is 1+1 >2? • Market sectors • Client types • Skills • Business Processes • Cultural alignment • Competitor analysis • Leverage • Diversity • Strategy and Business Plans • Risks and Opportunities.
4	**Indicative offer** Based on management accounts, prepare an indicative offer subject to due diligence and The Firm plc board approval.
5	**Heads of terms and due diligence** Agree heads of terms and commence due diligence.
6	**Sale and purchase deed** Prepare a detailed sale and purchase agreement.
7	**Completion and integration** Date for completion of the transaction and commencement of operations in accordance with the approved Integration Plan.

This chapter has highlighted many of the challenges with M&A either as the acquirer or the acquiree. To harness the synergies between the two firms, the integration challenges should not be underestimated.

8 Project management

> The definition of insanity is doing the same thing over and over and expecting different results.
>
> (Albert Einstein)

This chapter outlines the fundamentals of project management, leadership and governance for successful delivery of our projects, when measured against the project objectives. These measurements against the agreed project objectives are often referred to as key result areas (KRA). It includes the essence of collaborative contracting, establishing a project-management office, progress assessment and reporting, managing change and some suggested ingredients for project success.

With responsibility starting from top to bottom, the first section addresses project governance.

8.1 Project leadership and governance

The role of the Project Governor, Project Sponsor, a Governance Committee of unincorporated Design Joint Venture (DJV), or a collaborative contract is to operate as a Project Board performing all of the functions required to successfully deliver the project. The role is to undertake the strategy and governance functions. In this context strategy means the 'blue sky' vision for success and governance means the day-to-day review and oversight. The commercial interests of each participant are best served by meeting or exceeding the agreed common project objectives:

- Create an inspirational vision/purpose for the project
- Establish the principles and set challenging objectives
- Agree cost and other performance targets
- Set policy and delegations of authority
- Approve the project management plans
- Appoint and empower the Project Director
- Approve the appointment of the Project Management Team (PMT)
- High-level stakeholder interface

- Harness the best resources from the participant organisations
- Monitor team performance and take corrective action
- Resolve inter-participant conflict
- Provide appropriate oversight and intervene when required.

The role of the Project Management Team (PMT) is:

- Be responsible for the health and well-being of the wider project team
- Deliver outcomes that exceed the project objectives
- Appoint and empower the wider project team
- Day-to-day management of the project
- Provide effective leadership to the wider project team
- Measure, forecast and report performance to the Governance Committee
- Take appropriate corrective action.

The role of the Wider Project Team (WPT) is:

- United project team with one purpose and common project objectives
- Each position has clear accountability for outcomes
- Single person structure – no person-to-person marking
- All people appointed on a best-for-project basis
- No duplication of systems or processes
- Deliver the project as one team
- Develop and monitor risks and opportunities
- Create a high-performing team environment
- Ensure enhanced personnel and professional development while on the project.

Governance Committee

The Governance Committee will consist of senior representatives from each of the participants. Its primary role is to provide strategy and governance through appropriate supervision, guidance, oversight, coaching and mentoring to the project. Strategy is in terms of blue-sky thinking, a vision for success and pre-empting challenges. Governance is in terms of the day-to-day functions to ensure the project remains on-track.

The overriding function of the Governance Committee is to ensure that the project achieves its objectives and that the participants fulfil all their obligations, while also satisfying the corporate requirements and constraints of all the participants.

Consistent with the principle that the participants have a peer relationship where each has an equal say in decisions for the project, all decisions of the Governance Committee related to the project must be unanimous, or at least by consensus. For the Governance Committee to function effectively, the representatives need the right attributes, including:

- Superior leadership skills, including an ability to challenge their own pre-conceived ideas and unconscious biases and a commitment to further develop their leadership capabilities through the project
- Delegation of Authority (DoA) and the ability to make high-level Governance Committee decisions on behalf of their organisation
- A long-term perspective on the business aspirations and strategies of their respective organisations and a high regard for the relationships with the other participants
- Particular skills in areas that will add value to the project
- The ability to 'wear two hats' to ensure that both their organisation and the project achieve their objectives, and an ability and willingness to see things from each other's perspective.

The leadership factor

As in any successful organisation, the tone, drive and commitment must come from the top. The Governance Committee and the PMT each play a critical role in providing effective leadership to turn the aspirations and commitments of the participants into reality.

As the peak leadership body within the project, the primary role of the Governance Committee is to create and sustain an environment in which the project objectives can be met or exceeded. Similarly, the PMT, as a management team, must create and sustain an environment conducive to producing the results it is accountable for and committed to.

Influential leadership is the power of supportive relationships that make a positive difference, whether in a position of (direct) authority or as a team player at the coalface. The key word here will always be: Relationships – Relationships – Relationships. While individuals can achieve great things, well-led, or cohesive teams overcome immense obstacles, such as the first man on the Moon, or conquering Mt Everest. Such endeavours only succeed because of the effort we invest in developing and maintaining relationships, both internal and external. Particularly, seemingly small actions such as positive/constructive feedback, showing appreciation and recognition can have an effect far greater than the small effort it takes to make such gestures.

Although we may not all agree on some of the fundamental issues of business, the desire to understand the situation of others and build genuine trust through authentic empathy has always delivered the most enduring and powerful results.

Strategy and governance

The primary role of the Governance Committee is to balance strategy and governance. In this context strategy is defined as the 'blue sky' outlook, setting up the project for success. Governance is the day-to-day functions. The challenge for the Governance Committee is to establish the control distance to the PMT. While the Governance Committee is keen to empower the PMT, there needs to be appropriate governance and oversight, while being neither too distant nor micro-managing the PMT. There needs to be appropriate oversight, such that the Governance Committee can intervene when required, adding value of putting out spot fires. This bond of trust is discussed in more detail in the People and Culture section below.

The most effective Project Governors demonstrate and exude:

- Vision that inspires
- Passion that motivates
- Integrity that elevates
- Courage that emboldens
- Humility that connects.

An adage used by many Governance Committees is 'noses in, fingers out'. This means, monitoring the pulse of the project with effective engagement and empowering the PMT to get on with the project, with appropriate oversight. This does not mean hovering around at 30,000 feet. It does mean having the appropriate distance between the Governance Committee and the PMT, not unlike the relationship between a parent and their children based on trust and respect of the values instilled in them. However, if the project doesn't feel or smell right, then there is a time for intervention. Early intervention will often stop small issues 'snowballing' into much larger issues. With a diverse Governance Committee, each member normally brings a different level of KSEA to the project. Depending on the project and the issue at the time, one of the Governance Committee members should put their fingers in, get their hands dirty and help the project team address the challenges.

8.2 Project management

> If we fail to plan, then we plan to fail.
>
> (Benjamin Franklin)

While many firms have established practices, project management practices should be developed around the industry standard principles of ISO 21500 (International Standards Organisation, 2012), which are similar to other frameworks such as PMBOK (Project Management Institute, 1996), PRINCE2 (2016), P3M3 and Agile Project Management. The core functions of project management are:

- Safety and Ethics – including well-being and zero harm
- Integration – including interfaces, interactions, cooperation and coordination
- Client
- Stakeholders and community
- Scope
- Resourcing
- Schedule
- Commercial

- Technical quality
- Procurement
- Communications
- Risks and Opportunities.

Efficient and effective management of project time, costs and quality will yield benefits to our clients and shareholders in both financial and non-financial terms. Everyone profits from delivering on time, within budget and to the functionality requirements. Project management is seen as a key differentiator in a marketplace where technical excellence is often undifferentiated within the peer group. The ability to successfully plan and manage our projects provides the assurance sought by clients and line management, particularly as we embark on more innovative delivery methods. Recent industry experiences highlight there needs to be an immediate and committed response to develop a more disciplined certain and consistent approach to the planning and managing of our projects.

As a project-oriented, multi-dimensional business, the project-management philosophy should be structured around these core elements. Adapting the 4Ps of marketing (Kotler, 2012) outlined earlier, with a fifth P being Program:

- People: The right project manager is appointed for the task, accountable to a Project Governor who will actively hold the project manager to account. Appointment of key personnel with key roles and responsibilities
- Process: An agreed project management plan is prepared based on the proposal work plan, which is implemented and reported against regularly
- Product: Assurance of the quality of the deliverables and progress through appropriate and timely reviews using lead and lag metrics
- Price: Execution with the requirements of cost, schedule and functionality within the budget or target estimate
- Program: Scheduling activities in a logical and sequential manner with identification of key milestones.

Based on the research and experiences the following twelve initiatives will assist in developing the PM competencies and supporting tools:

- Implement a project-management policy and practices and apply them consistently
- As a project-oriented firm, implement a philosophy in the organisation that drives a project-management approach for both internal and external clients
- Review the project-management learning and development and career-development pathways
- Develop a culture of open communication with no surprises and sanctuary for early notice
- Select the right person as the project manager every time and provide appropriate supervision and support

- Ensure a fit-for-purpose project management plan is developed based on the work plan in the proposal/contract and implement the plan with regular reviews
- Focus on the deliverables (outputs) and milestones, not just the resources (inputs)
- Define the responsibilities of the project manager, the accountabilities of the project governor and the stakeholders to be consulted/informed using a RACI matrix
- Share project success with the entire project team and line management
- Use industry project management tools and systems as support, not as a substitute for effective planning and managing
- Measure progress regularly with weekly updates and monthly reports as a declaration of the project performance. Evaluate the earned value monthly and determine the forecast to completion
- Develop the emotional intelligence (confidence, interpersonal skills, resilience, courage, charisma and character) of our project managers.

As a project-oriented firm the way we plan and manage our projects provides the assurance to our clients and shareholders (via line management) that our projects will achieve or exceed their objectives.

8.3 Collaborative relationships

As discussed previously, effective collaboration between enlightened Project Owners, partners, consultants, contractors and related stakeholders are based on trust. Trust-based relationships are earned by working closely together to develop winning solutions and deliver projects successfully. And we should share in that success through enhanced gross margins and repeat business with reduced cost of selling. If we are working together for the first time, developing trust-based relationships takes time and investment. Our greatest 'gift' to build trust is our time and knowledge.

Alliance contracting is a form of collaborative contracting (Department of Infrastructure and Regional Government, 2015). An emerging form of delivery is Integrated Project Delivery (IPD) (American Institute of Architects, 2017) as a project delivery approach that integrates people, systems, business structures and practices into a process that collaboratively harnesses the talents and insights of all participants to optimise project results, increase value to the Project Owner, reduce waste, and maximise efficiency through all phases of design, fabrication and construction.

IPD principles can be applied to a variety of contractual arrangements and IPD teams can include members well beyond the basic triad of the Project Owner, designer, contractor and operator. In all cases, integrated projects are uniquely distinguished by highly effective collaboration among the Owner, the prime designer, and the prime constructor, commencing at early design and continuing through to project handover.

Again using David Maister's trust equation outlined above (Maister, 2001), adapted to this environment, the following triptych (see Table 8.1) provides further explanation of what this means to us for collaborative contracting. This framework endeavours to align the hard iron-triangle of time, cost, quality with the softer relationship elements that are arguably the ingredients to achieve project outcomes.

8.4 Project Management Office

A Project Management Office (PMO) is the repository for the project-management practices and considers how to continually improve the planning and managing of our projects. The purpose of the PMO is to provide a common platform of project-management systems and processes to facilitate project success. The objectives of the PMO are:

- Produce best-practice templates, processes and guidelines
- Project control systems and tools for better project reporting and consistent management
- Knowledge management including feedback and continual improvement
- Training and development and core competencies of project managers.

This section provides a first pass of what should be included in the PMO 'white van'. The electronic 'white van' includes templates, examples and guidelines based on past experiences and industry standards. These should include:

- Default procedures and systems
- Suggested requirements, guidelines and practice notes
- Mandatory requirements.

To set up a project for success, the 'white van' is dispatched and the SWAT team of experienced personnel support the project team with the implementation of the appropriate project controls and template documentation. The indicative elements of a PMO are outlined in Table 8.2.

A PMO is not a Business Management System (BMS). The BMS defines the processes for managing the business. The PMO is a knowledge centre for managing our projects and needs to be adaptable to the Owner's requirements and when we joint venture with allied design partners.

8.5 Progress assessment

Earned Value Measurement (EVM) is a project-management technique for measuring project performance and progress of the project measured against:

- Deliverables – design documentation (drawings, reports, technical specifications)
- Schedule – the schedule performance index (SPI), which is Earned/Planned
- Costs – the cost performance index (CPI), which is Earned/Spent.

Table 8.1 Elements of trust-based relationships

Credibility	*Reliability*	*Intimacy*
Credibility means that we have a proven track record and we have 'done this before'.	**Reliability** means we always respond on time (or advise otherwise with due notice).	**Intimacy** means we understand the psychology of the relationship and harness our emotional intelligence.
Care means we honour the standard of care required by the end Owner.	**Responsiveness** means we are pre-emptive, predictive and intuitive to the project challenges.	**Innovation** means challenging existing paradigms and establishing new methods and approaches for better than business-as-usual outcomes.
Compliance means we provide assurance of compliance with the end Owner's project requirements.	**Reliance** means that the D&C client can rely on our advice for expeditious approval and construction and we warrant that advice under contract.	**Imaginative** means thinking outside-the-box in developing value-for-money solutions.
Competency means the key personnel have the requisite competency and experience.	**Representation** means that we have properly represented our capability and resourcing per the contract and the CCA.	**Interest** means we have a best-for-project (B4P) approach balancing our self-interests with being servile.
Capability means the technical capability of the entire firm sourced locally, nationally and internationally or with specialist consultants.	**Relationships** mean we understand and empathise with each other's perspectives. We can stand in each other's shoes.	**Integrity** means values-based assurance is our moral compass. Every behaviour is with integrity that does not compromise our ethics or reputation. Integrity is the internal sat-nav that guides all project personnel to do what is right.
Constructability means we consider safety in design and we seek to understand how our designs will be constructed.	**Responsible** means we accept responsibility for our actions with clarity of scope and are accountable for the outcomes.	**Informed** means we understand all the project issues, sourced from the D&C client, end Owner and other related stakeholders.
Client centricity means we appreciate the difference between the end Owner's 'needs' and the D&C client's 'wants'.	**Risk** means uncertainty and we are risk-savvy in the identification, assessment and mitigation/management of all project risks. We intervene when required.	**Incentivisation** means we are prepared to put margin at-risk to share in better project outcomes.

(Continued)

Table 8.1 (Continued)

Credibility	*Reliability*	*Intimacy*
Commitment means we are committed to achieving the design milestones.	**Risk Acumen** means we understand the human factors in developing a risk-based culture in the project management and governance functions as the key ingredients to the assurance of project success.	**Indicators** means we embrace the project objectives and use the KPIs as a true measure of our performance against agreed metrics.
Change means we understand the difference between design development (up to stage 1 submission) and design change using a VE approach.		**Improvement** means harnessing hard-earned lessons learned in the aspiration of continuous improvement and enhanced productivity.
Contingency means considering inherent and contingent hours (aka design costs) for the known unknowns and the unknown unknowns.		**Inquiring** means questioning as the antidote to (implicit) assumptions that so often incubate mistakes.
Confidence means we are prepared to back ourselves to achieve better than business-as-usual outcomes.		**Independence** means we embrace independent peer review and catalyst teams with a fresh-eyes perspective.
Communication means '*we talk early, talk little and talk often*', particularly in periods of uncertainty. We communicate effectively through verbal, non-verbal and written communications. We page-turn our deliverables.		
Controls means we balance empowerment with appropriate controls, with a just culture.		
Culture means we provide active project leadership to develop a positive and engaging project culture, often co-located in a dedicated project office.		

Table 8.2 Indicative elements of a PMO by service type

ISO 21500 Function	*Consulting*	*Design*	*PM*	*Construction Services*	*Support Services*
Aka	Advisory		PMC, Project Management Contract Administration, EPCM	Construction Management, D&C, EPC	O&M
Objectives	Stated project objectives PD governance	Stated project objectives Success models KPIs Governance Committee governance	Stated project objectives Success models KPIs PD governance	Stated project objectives Success models KPIs Governance Committee	Stated project objectives Success models KPIs Governance Committee
Health, Safety & Environment	Health & Safety Management Plan (HESP) Travel journey plans Subcontractor Principal Contractor	Safety in Design Reviews (SIDR) Health & Safety Management Plan (HESP) Narrative about collocated project office environment	Safety Management Plan Key lead and lag indicators	Principal Contractor obligations Safety Management System (iSMS) Environment Sustainability tool (INVEST)	Principal Contractor obligations Safety Management System (iSMS)
Scope	PM Plan with discrete tasks and milestones Scope change	Engineering & Design Management Plan (E&DMP) Scope & Technical Requirements compliance Stage 1, 2 and IFC gates Owner's review and comment/approval process	Project Management Plan (PMP) Management Plans Scope & Technical Requirements	Project Execution Plan (PEP) Management Plans Scope & Technical Requirements	Project Execution Plan (PEP) Management Plans

(Continued)

Table 8.2 (Continued)

ISO 21500 Function	*Consulting*	*Design*	*PM*	*Construction Services*	*Support Services*
		Design packages IFC/CPS demarcation Design change process (design development v. design change) Quantity growth			
Schedule	Design programme using XLS	Design programme using MS Project or P6	Project programme using P6 Float management	Construction programme using P6 Milestone reporting Float management	Contract term programme using P6
Cost	Progress measurement v actual (accrued) costs Forecast at completion	Earned Value reporting of design progress including Planned Value (PV), Earned Value (EV) by deliverables, Actual Cost incurred (including accrued costs) and assessment of SPI= EV/PV and CPI= EV/AC Forecast at completion Reporting of CPS expenditure	Cost plan EV based on installed costs Forecast at completion	Cost plan and cost coding Cost control (PRISM) EV based on installed costs Forecast at completion Cash flow	Cost plan and cost coding Cost control (PRISM) EV based on installed costs Forecast at completion Cash flow
Quality	Technical reviews as per BMS	Strategic project reviews	QA reviews Document control (e.g. Aconex)	QA/QC requirements	QA/QC requirements (Document control (e.g. Aconex)

		Technical reviews as per BMS Stakeholder approval process/workflow Narrative about updating drawings with process for hold clouds and update clouds Document control, filing and archiving (Project Wise)		Document control (e.g. Aconex) Completions assurance	Completions assurance
Contracts	Contractual compliance Variation management	Key terms of the head contract client management Sub-consultant management Planning for claims Variation management	Key terms of the head contract client management Superintendent functions Contract administration Notices	Key terms of the head contract client management Subcontract management Variation management EOT process and notices Procurement register, purchase orders Subcontract register, subcontract terms Contract Management System (Primavera) Notices	Key terms of the head contract client management Subcontract management
Resourcing	Resource planning	Design Org Chart Resourcing Plan	Org chart with key personnel	Org chart with key personnel	Org chart with key personnel

(Continued)

Table 8.2 (Continued)

ISO 21500 Function	*Consulting*	*Design*	*PM*	*Construction Services*	*Support Services*
		Narrative about matrix of Design Leads and Area Managers Position descriptions for key personnel Reward and recognition examples	Position descriptions	Project team development Position descriptions Industrial Relations Readiness for work (certifications)	Project team development Position descriptions Industrial Relations Readiness for work (certifications)
Communications	Weekly updates Monthly progress reporting	Weekly updates Monthly progress reporting	Weekly updates Monthly progress reporting	Weekly updates Monthly progress reporting	Weekly updates Monthly progress reporting
Interfaces	Interdisciplinary Coordination (IDC) Collaboration with sub-consultants	Interdisciplinary Coordination (IDC) Collaboration with Other Consultants Work packaging providers (Globalshare) Stakeholder approvals	Stakeholder & community plan Project contractors Asset owners and operators	Delivery partners Stakeholder & community plan Asset owners and operators	Delivery partners Stakeholder & community plan Asset owners and operators
Risk & Opportunities	Qualitative risk assessment	Project Risk Management Plan (PRMP) Risk register identification and administration (iRMS) Value engineering Qualitative/quantitative risk assessment	Project Risk Management Plan (PRMP) Risk register identification and administration (iRMS) Value management Quantitative risk assessment	Project Risk Management Plan (PRMP) Risk register identification and administration (iRMS) Project optimisation	Project Risk Management Plan (PRMP) Risk register identification and administration (iRMS) Contract optimisation
Knowledge	Filing, archiving Lessons learned	Filing, archiving Lessons learned	Filing, archiving Lessons learned	Filing, archiving Lessons learned	Filing, archiving Lessons learned

EVM of progress is somewhat subjective even on well-managed projects. AS4817 (Australian Standards, 2006) provides the methodology for determining progress comparing:

- Planned (aka original or revised budget)
- Earned
- Spent (aka actual costs).

Indicative progress curves are shown in Figure 8.1 for the Planned, Earned and Spent.

It is recognised that assessment of progress in the first 20 per cent is the most uncertain. However, it is this period that is most critical. The 20 per cent milestone is the first real opportunity to forecast the likely outturn. This involves forecasting:

- Estimate to Completion (ETC)
- Estimate at Completion (EAC).

All projects will progress through some form of progress curve (aka s-curve) and experiences suggest that 'what gets measured gets managed' (Collins, 2001). The more reliable the data, the better the decision making at critical stages of the project.

For design projects, the most reliable EVM is by using the Design Package Register (DPR) and amortising the Original Budget across the packages based on the effort required. The common milestones for design documentation are:

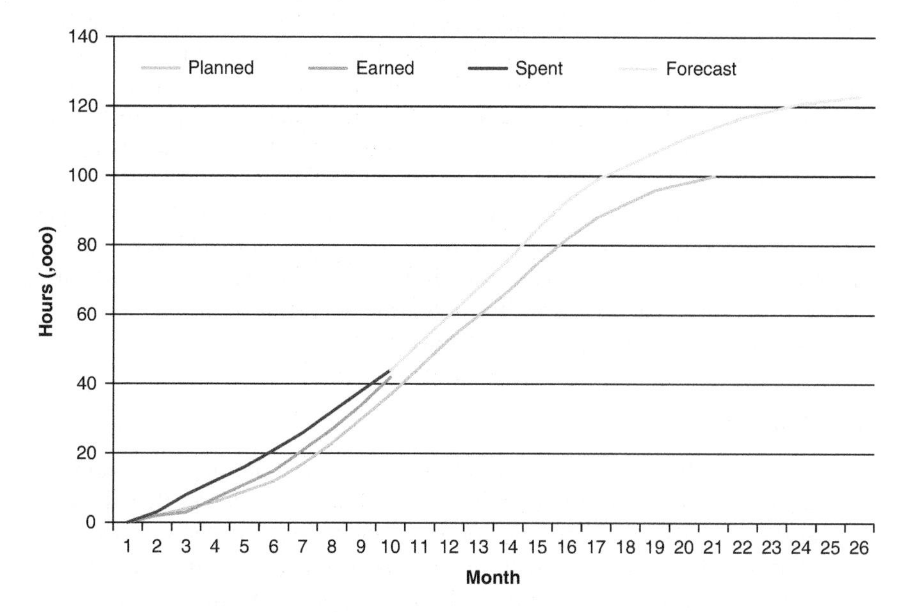

Figure 8.1 Indicative progress assessment

For infrastructure projects:

- Preliminary Design – internal review 30% and external submissions 35%
- Detailed Design – internal review 80% and external submissions 85%
- Final Design – internal review 95% and external submissions 100%.

For building projects:

- Concept Design – 15%
- Schematic Design – 35%
- Detailed Design – 90%
- Final Design – 100%.

It is common to assess progress within these milestones to provide a reliable evaluation of progress and hence payment of work performed.

Complexity arises when there is significant design change and the Planned curve needs to be adjusted for approved variations and professional judgement is required for recognising pending un-approved variations. If there is significant change and it is uncertain what the Revised Budget is, then it is suggested that the earned value be re-assessed based on the Actual Costs spent and re-assessment of the ETC. That is:

$$AC + ETC = EAC \text{ and then } EV = AC/EAC$$

This highlights the need for detailed scope definition in the proposal phase and a clear process for change management, or even better instil a policy of NO CHANGE. There is an old adage in D&C contracting, 'design as tendered, then construct as designed'. Design change can be like a virus, particularly if it is late in the design process. It should be absolutely clear the design development ends at the Stage 1 submission. After that it is a design change. This is discussed in more detail in this next section.

Many of our projects have suffered with delayed completion of the final 20 per cent, known as the 'sting in the tail' of the progress curve, with additional detailing and incorporation of comments from the Owner, stakeholders and third-party reviewers. Closing out comments and soliciting approvals from third parties is often the most challenging part of the project.

8.6 Managing change

As highlighted above, for many consulting projects the outturn scope and fee is often significantly different from that tendered. This reinforces that the purpose of the tender is often for selection of the preferred tenderer, not necessarily performing the work. For design projects in D&C projects, the challenge is to differentiate design change from design development. In this context design development means developing the tendered design in the first stage to enable cost planning, constructability assessment, value engineering, and 'freeze' the design to then document the design with certainty.

We need to balance the client's preferential wants with their defined needs as specified in the contract. Preferential and pedantic interpretation of the project requirements by the client and their technical advisors and asset betterment from key stakeholders can lead to significant scope creep.

All projects will face challenges and we need to be prepared to have senior representative negotiations with the client. Our senior representatives need to be commercially savvy and have the guile and street-smarts to be able to negotiate fair and reasonable outcomes, in good faith.

For both consulting and design projects, the onus is on the consultant to demonstrate the causation of the change and the impact. The doctrine of law is that the burden of proof lies with the party making the allegation – 'He/She who alleges, must prove'.

Consulting projects

Changes can be small-scope changes, or larger contractual claims. If the latter, a fully particularised claim for additional services has the core elements listed in Table 8.3.

If the claim is likely to be contested by the client and therefore progress through the dispute-resolution process, the claim needs to be well-prepared with objective evidence to support each of these elements. This highlights the need for thorough contract administration and project documentation.

Table 8.3 Indicative elements of a claim for additional services

Element	*Description*	*Causation*
1. Basis of Claim	What was the causation of the claim?	What
2. Entitlement	Pursuant to the relevant clauses of the contract, the consultant is entitled to reimbursement of the additional services.	Why
3. Notification	Refer notices provided and received on [date] instructing the consultant to change the design.	Who and When
4. Impact	How did the change impact the performance of the services? Impact on critical path activities? Re-sequencing? Delay and disruption? Timing of the change?	How
5. Evidence	Multiple revision of reports, drawings, etc.	Proof/Evidence
6. Quantification	Indicative fee adjustments provided, but with no agreement on the fee for the Delivery Phase, services were provided on a Time and Expense (T&E) basis.	Consequence

Design projects

The principles for the provision of design services for fast-tracked D&C contracting are:

- We conduct the design services with the mantra 'design as tendered, construct as designed'
- Design development (involving cost planning, constructability assessment, etc.) is frozen at Stage 1
- Time is of the essence. We only deviate from our path to success if we have a written instruction from our D&C Contractor client (based on a considered VE decision to deviate from the original plan)
- Untimely change disrupts our workflows and productivity and can have viral consequences. The knock-on effects of a change must be considered and the D&C Contractor advised to assist in the VE decision making
- We are liable for delay. We will only change if we are provided with time relief and reimbursement of the additional costs, pursuant to the contract
- We limit the number of contractual Milestones (aka target dates) and provide timely notification of EOT when delayed
- After Stage 1, we monitor each and every design change in the Design Change Notice (DCN) (whether or not they constitute a variation), to be included in our monthly reporting. The Design Change Register (DCR) is the registering of all the DCNs.
- The DMP clearly defines the design change process, to be embedded in the contract
- We strictly administer the D&C Contractor's contract provisions in a highly professional manner. The role of the commercial manager is to implement the project controls and fulfil the contract administration requirements.

To demonstrate the demarcation between design development and design change, the framework in Table 8.4 endeavours to provide the distinction, to be articulated in the contract or the accompanying management plan.

8.7 Progress reporting

Anecdotally there is sometimes a perception that robust contract management and administration does not align with being a client-centric firm and will impact the client relationships. Being professional and looking after our interests and that of the client should result in win/win outcomes. We serve our clients professionally, we should not be servile or subservient. Industry surveys clearly show that our clients expect better service delivery from our industry.

As explained in Chapter 4, the Beaton survey (Beaton Research and Consulting, 2017) of consulting engineering firms, highlights that clients highly value the following.

Table 8.4 Design development versus design change

Stage	*Description*	*Purpose*	*Design development*	*Design change*	*Indicative triggers for design change*
1	Preliminary Design	Development of the design to confirm compliance with the Concept Design for review and approval by internal and external parties	Changes to Concept Design as required to achieve the Preliminary Design. Examples: Changes within Design Packages necessary to align with other design areas. Inter-disciplinary coordination. Changes to Concept Design as required to confirm compliance with the OPR.	Any material change to the Concept Design concepts or design intent. Material change means a significant amount of rework or additional design effort and change to the programme milestones. Examples: Creation of new scope (Design Package) Splitting a design package to expedite construction (aka acceleration cost) Late change (within 2 weeks of Stage 1 baseline submission deadline) to Design Package requiring rework of deliverables. Other	Constructability Client change
2	Detailed or Certified Design	Documentation of the design for review and approval by internal and external parties; and tendering and procurement	Nil - all change post Preliminary Design submission is Design Change.	Any change to the Preliminary Design causing rework or additional design effort and change to the Programme Milestones	Constructability Cost planning Value Engineering Subcontractor request Approvers request Client change Stakeholder request
3	Final Design	Finalisation of the Design Documentation for construction purposes	Nil – all change post Certified Design submission is Design Change.	Any change to the Certified Design causing rework or additional design effort and change to the Programme Milestones	Constructability Approvers request Client change Stakeholder request

- Reliability: deliver on-time and on-budget, say what we'll do and do what we say. Clients can be very demanding and we must supply the deliverables expeditiously. It is simple economics: supply must match demand
- Responsiveness: we should openly communicate with a no-surprises approach. Generally clients can work-around late deliverables provided we have given them sufficient notice. We should be proactive in keeping them up-to-date and report to them when we say we will. We should also bring ideas, insights and innovations
- Quality documentation: the deliverables (drawing, reports and specifications) must be complete and correct, suitable for their purposes. Not over-detailed and without omissions or defects and to the industry standard level of detailing.

Progress reporting should be a combination of weekly updates and monthly reports. Weekly updates include:

- Achievements in the previous week
- Plan for the next week
- Look-ahead and key issues.

Monthly reports should be well-articulated and tailored to meet the needs of the client, for their communication upstream. While many reports will be based on the lag project and financial data, the narrative needs to explain the meaning behind the data. A well-written narrative provides the project story that supports the data and trends. Dashboard reporting is useful, but is not a substitute for a well-written report.

Progress reports also provide a chronology of the project issues, in the defence of any claims, and so they should be diligently prepared, with a factual record and statement of key issues.

8.8 Project success

This project management website (Project-Management.Com, 2016) identified the top ten causes of project failure. A more positive perspective would be to consider these as a contra proposition.

Poor preparation

We need to have a clear picture of what we're going to do, in advance – as much as possible. Otherwise, we may find ourselves 'up stream without a paddle'. We need to know what project success looks like at the beginning and don't lose focus of it. Hence, if we don't have a clear focus at the earliest stage of the process, we are making things harder on ourselves. Have a meeting, even if it is lengthy, with stakeholders to discuss their expectations on cost, time and product quality. Know how we will execute our tasks in order to meet everyone's expectations.

Inadequate documentation and tracking

This is the responsibility of the PM. Tracking milestones is how we are going to know whether we are meeting expectations. Proper recording and monitoring lets the PM identify where more resources are needed to complete a project on time.

Bad leadership

When we see this word, leader, we usually think, the PM. However, the people at each management level have a responsibility to ensure that the project is successful. Management should not micromanage but should provide appropriate oversight, supervision and support to ensure that the PM can follow through with the expectations placed upon them. There is an appropriate balance between empowerment and controls.

Failure to define parameters and enforce them

When you're a PM, it's imperative that you're able to work well with your team. If and when tasks or goals are not met to standard, there should be ramifications. Rank tasks by priority and assign them to the most proficient individual.

Inexperienced project managers

A PM has a lot of responsibility. We need to assign people to management roles that have matching knowledge, skills, experience and attributes (KSEA). In some cases, and perhaps more often than not, inexperienced managers are given projects. They may be very capable of managing projects, but the key is to keep them at a level where they can succeed. Otherwise, you will set them up for failure. On the other hand, there's nothing wrong with a challenge, just don't make it beyond their reach or thow them in the deep end without support.

Inaccurate cost estimations

There may be times when your cost estimates are completely off. As you know, when resources run out, the project stops. Prevent this by identifying the lack of resources early on.

Little communication at every level of management

Whether it's between upper management, middle or with the team, it's disastrous to have poor communication. Everyone should feel free to come forward to express their concern or give suggestions. When everyone is on the same page and there's transparency, workflow is at an optimum level.

Culture or ethical misalignment

Company culture must be comprised of competence, pro-activeness, and professionalism. If it isn't, team members will not be motivated to do their best. Basically, everyone involved must be invested in their part of the project to successfully complete it.

Competing priorities

When there are not enough resources, there's bound to be competition between personnel resources and funding. Having good cost estimations at the start will eliminate this problem.

Disregarding project warning signs

When a project is on the verge of failing, there will always have been warning signs. Early warning is key, and corrective action can then be taken. Provide a culture for early warnings and no surprises. Taking action immediately with experienced staff can save the project. Otherwise, the whole endeavour can go down the drain.

9 Operational management

This chapter describes the operational management activities of a large professional services firm. It addresses some of the issues with corporate governance of a subsidiary firm of an international organisation and how to manage in uncertain times, whether a heated or depressed market.

9.1 Corporate governance

With the internationalisation of many professional services firms, many are listed on the LSX, NYSE and other international stock exchanges with the local entity a wholly owned subsidiary of the parent company. This section provides the framework for corporate governance by the Board of Directors of the subsidiary company outlining the processes of the statutory board for managing corporate risk.

There are two appropriate sources of corporate governance.

In Australia best practice is represented in the ASX guidelines (ASX Corporate Governance Council, 2010). These guidelines apply to Australian listed companies by virtue of the ASX Listing Rules. For unlisted Australian companies, the guidelines represent best practice.

The requirements of the ASX guidelines and the UK Combined Code (UK Corporate Governance Code, 2017) are similar in many respects. Both acknowledge that the application of corporate governance must suit the company and add value. This flexibility of implementation means that it is not necessary to adopt either the ASX Guidelines or the UK Combined Code but rather we can adopt the best practice of each system to the extent that it is relevant to the firm.

Charter

The directors of the subsidiary firm should decree that the role of the local Board shall be to:

- Lay the foundation for management and oversight
- Structure the Board to add value
- Promote ethical and responsible decision making
- Safeguard integrity in financial reporting

- Respect the rights of the shareholder and actively participate in its corporate governance processes
- Recognise and manage risk
- Encourage enhanced performance
- Remunerate fairly and responsibly
- Recognise the legitimate interests of stakeholders.

Statutory responsibilities – minimum standards

The firm should affirm its commitment to comply with all law and regulations including the requirements of the Corporations Act.

Although there is no reference in the Corporations Act directly to corporate governance, the Corporations Act does lay down a series of mandatory minimum rules and standards for the conduct of the Company and its Directors which represent the minimum mandatory corporate governance for the firm. The relevant provisions in the Corporations Act include:

- Replaceable Rule 198A which empowers Directors to manage the affairs of the Company
- The various Directors' duties to act in good faith for a proper purpose (s181) and the avoidance of conflicts of interest (s182 and 183), to exercise care and diligence (s180) and the Financial benefit provisions in Chapter 2E
- The provisions allowing Directors to delegate their powers subject to supervision (s189, and 190)
- Ensure the Board acts in the best interests of its parent
- The provisions regarding insider trading (Chapter 7), insolvent trading (s259G), and the holding of meetings (Part 2G).

Appointment of Directors and Chairperson

The role of the subsidiary Board is to provide the strategic direction for the Region and assure the parent shareholder of the diligent and compliant administration of the company in accordance with all relevant statutory obligations.

For the appointment of new Directors, the Managing Director shall provide a recommendation to the Board. A quorum for a statutory Board meeting shall require the attendance of at least three (3) directors. Directors shall use every endeavour to attend statutory Board meetings or dial in by phone. Every decision, determination or resolution of the Board shall be made unanimously on a best-for-business basis.

Roles and responsibilities of Directors

Directors have a duty to independently consider each issue that comes before the Board and to ensure that all important issues are raised. Directors are normally also responsible for line management or functional directorates. Directors shall ensure that there is clear demarcation between management responsibilities and director obligations so there is no conflict of interest. If a perceived or actual conflict arises, the Director shall abstain from any decision making.

The firm, as a private company is not required to have independent or non-executive directors on its Board. However the Board will consider whether the inclusion of independent directors and non-executive directors will add value to the Board on a case-by-case basis. This may be considered as an advisory Board or committee.

Role and responsibility of the Chairperson

Often the role of the Chairperson is held by the Managing Director and this could cause a conflict of interest. This is due to the private nature of the company and the oversight provided by its parent. The parent should recognise that best practice is to separate the roles and that ideally the Chairman should be a non-executive independent Director. The role of the Chairperson is to conduct the Board meetings in accordance with the charter.

Role and responsibility of external advisers

Instead of non-executive and independent Directors, the firm may appoint a number of advisors to the Board. The role of these advisors is to play a key role in bringing fresh ideas and assisting to formulate strategy.

Role and responsibility of the Company Secretary

The role of the Company Secretary is to assist the Chair in organising statutory and board meetings and monitoring the corporate governance of the organisation.

Director and Officers insurance

In order to protect the Directors and Officers of the company, the firm should maintain an adequate D&O insurance as well as a Statutory Indemnity policy.

Conduct of the meetings

Meetings shall be arranged by the Chairperson on a quarterly basis. Minutes shall be prepared by the company secretary and distributed to all directors within seven (7) days of the meeting.

Ethics and code of conduct

The directors of the firm will be required to:

- Act openly and honestly
- Operate a spirit of transparency
- Consider the interest of stakeholders (shareholders, clients, staff and third parties) in all decisions

- Not engage in any conduct in breach of any law
- Engage in frank debate
- Endeavour to do the right thing
- Bring to the attention of all directors' material issues.

Sub-committees

The Board may elect to establish sub-committees, such as:

Audit

The Board shall establish an audit committee with a separate charter. The Audit committee shall consist of three Directors including the Managing Director and excluding the Finance Director. The audit committee shall meet with the auditors as it considers necessary and shall consider financial information being provided by the Company to its parent.

Risk

The Board shall establish a risk committee consisting of three Directors plus relevant managers involved in risk prevention. The Board shall develop a charter and sit quarterly to consider specific risks relating to the company.

Other

Other committees may be established at the discretion of the Board to address critical issues.

Disclosure of interests

The Company Secretary will keep a standing list of material disclosures that is available for the Board. Each Director shall fully disclose any actual or perceived personal conflict of interest.

Best practice

The corporate governance policy should be reviewed on an annual basis to ensure that the firm's corporate governance practices keep pace with the firm's development.

9.2 Operational management

Most large professional services firms operate some form of matrix organisation. The key dimensions are as follows and the challenge is how to manage a multi-dimensional business:

- Profitability
- Sales Revenue
- Technical disciplines, capability centres or service lines
- Clients, types and market sectors
- Geography.

However, most organograms are a two-dimensional response to these multi-dimensional objectives and therein the poor siblings are often neglected. Again, what gets measured gets managed (Collins, 2001). Add to this the complexity of the interaction with the functional groups such as Marketing, HR, Finance, IT, QA, Risk, Strategic Initiatives (Maister, 2006).

As professional services firms endeavour to transition from selling technical services to being more client-centric, often the lead axis (solid line reporting) is clients, market sectors representing the sales revenue. The other axis is then business units, responsible for profitability (dotted-line reporting). Or vice versa. This structure often means that technical disciplines, communities of practice or technical capability groups are not given the attention they deserve to ensure they remain best-in-class.

No sooner than the organogram is drawn on a whiteboard, then silos, tribes, barriers and interfaces are formed. The successful firm is able to 'work the matrix' and break down these barriers through inclusiveness, integration and cross-communication. It is the organisational leaders who can work the matrix, which will harness the energy, enthusiasm and efforts of a multi-discipline, multi-geography, multi-sector professional services firm. McKinsey & Company describe this approach as working beyond the matrix organisation (McKinsey & Company, Tom Peters, 1979).

As explained earlier, professional services firms are often known for either their client centricity, operational efficiency, or technical excellence. It is unlikely that a firm will be outstanding in all three areas and often the focus depends on the focus and attention of the CEO.

Frank Staziowski of PSMJ (2016), suggested that professional services firms can be categorised as partnerships, businesses or a combination and this defines their approach to the three dimensions and the brand of the firm. As discussed in Chapter 4, the brand of the firm is a statement of what is the firm is known for. The operations of the firm must therefore align with that promise. Figure 9.1 describes the partnership versus business organisational interface matrix.

Most professional services firms operate along P&L operational centres, with teams or sections of capability groups. Operational Managers are therefore most focussed on:

- Utilisation (aka billability) of their staff
- Project performance and profitability (write-off/up)
- Working capital, being unbilled work-in-progress and debtor collections
- Sales and backlog
- People issues.

	Partnership	Business
Partnership	Partnership Centred Partnership (PCP) e.g. Boutique specialist firm	Partnership Centred Business (PCB) e.g. International design practice
Business	Business Centred Partnership (BCP) e.g. Global advisory firm	Business Centred Business (BCB) e.g. Large listed E&C firm

Figure 9.1 Partnership versus business centricity

Successful operational managers are people managers and recognise that in fact fully motivated and energised people will find the work, win the work, do good work and get repeat commissions. However, operational managers are often 'dumbed down' to simply 'workbook' managers focussing on the lag data, which can become highly demotivating to the personnel generating the revenue. That is the role of a commercial manager and that role should be more of a business advisor rather than score keeper. Astute operational managers are able to balance the lead with the lag data to effectively manage their teams.

9.3 Managing with uncertainty

The building and construction market is often cyclical. The Global Financial Crisis (GFC) of 2009 had a significant impact on the infrastructure sector in all geographies.

In recent years, the resources sector has remained flat after a decade of boom. Conversely, with the political imperatives many governments funded transport, water and social infrastructure projects through asset sales and 'good debt' borrowings. This section provides some fundamental principles for managing during an economic downturn, such that we can emerge as one of the leaders in each of our sectors.

While we are generally confident of the longer term outlook for the infrastructure sector as outlined in the introductory sections, the successful firm surviving the downturn will ensure success in the longer term. Manage in the downturn – plan for the upturn. What we learned from the GFC is that when dealing with a downturn, the ten fundamental principles for managing the firm when supply exceeds demand are as follows.

Back to basics

- Micro-manage weighted prospects, winning proposals, project performance
- Improve productivity of the work in hand
- Work harder, work longer
- Critically review the investment in non-project time.

Act decisively

- Demonstrate leadership – provide certainty in uncertain times
- Have the courage to make the right decisions
- Enhance operational effectiveness and streamline business processes.

Working capital

- Backlog is vital and cash is king!
- Expect clients will want to defer payments. Chase debt
- Fill up the backlog with quality sales
- Review the backlog, expect projects to be cancelled or further trade-ups deferred.

Focus on the important (not the urgent, not the self-interest)

- Right people, doing the right things at the right time
- Focus on the short term, while thinking of the longer term
- Focus on the task, not the role.

Match costs to revenue

- Income will be depressed – match overheads to the income stream, planned and actual
- Transition fixed costs to variable. Outsource non-core functions
- Use more sub-consultants, contract staff.

Reliable information is key to success

- Capture real-time intelligence from our target clients
- Frontline personnel must provide reliable client intelligence
- Update the weighted prospects and pipeline weekly
- Recognise that clients' perception of value will change.

Plan for different scenarios

- Balance strategy with opportunity – plan for the worst, likely and best scenarios
- Reduce horizons – strategise monthly, not annually
- The workload will be lumpier, plan for success or otherwise.

Recognise the value of key people

- Retain the right people who will grow the business after a downturn
- Maintain confidence through regular communication
- Review incentives and protect the future leaders of the business.

Seize opportunities

- Opportunities will be fewer and more fiercely fought
- Secure most win opportunities
- Identify quick wins to infill workload.

A more competitive environment

- Protect our target clients – they will be under siege from competitors
- Clients' buying behaviours will change
- Tier #2 competition will increase, driving down prices
- Stay close to our clients – get even closer to our competitors.

Conversely, when dealing with the upturn and a heated market, the ten fundamental principles for the operational manager for surviving and thriving in uncertainty when demand exceeds supply are as follows. Experience suggests that more businesses become insolvent during boom times, given their lack of focus on some of these principles.

- What we don't know may be more important than what we do know
- Sooner or later, something significant in your project will change
- Find the unexpected before it derails your project or business unit
- Sometimes a contrarian voice is needed
- Find the right balance between governance and management
- Find the right balance between empowerment and controls
- The need to provide Independent Assurance and 'correct weight'
- Innovation is the heart of risk mitigation
- Risks must be taken to seize opportunities
- The rear-view mirror doesn't help much when you're driving forward.

10 Commercials

This chapter provides an overview of the commercial management and financial operations of a large professional services firm to ensure there is sufficient working capital to fund the strategic objectives. For most professional services firms, the balance sheet comprises the net tangible assets (NTA) of cash, work-in-progress (WIP) and debtors, less creditors. In the current era, most professional services firms would elect to lease their fixed assets such as property, IT and other equipment. Therefore, solvency of the firm and the ability to fund strategic investments is determined by the working capital.

10.1 Profitability

Table 10.1 provides a listing of an indicative chart of accounts as recorded in the firm's financial system. All figures are derived on an accruals (not cash) basis recorded within the accounting period.

Salary multipliers

Most professional services firms use a salary multiplier as a primary measure of the cost of doing business. Some firms will add overheads as a percentage of the salary. Others amortise overheads on a per-head basis, others on an hourly basis.

For cost-reimbursable contracts, many clients will prefer to reimburse the consultant on a salary multiplier rather than a schedule of rates. This means that individual staff salaries and employee entitlements will be provided. A compromise to this 'open-book' approach may be the application of a schedule of rates that may be subject to independent financial audit to confirm alignment with actual salaries paid.

In collaborative contracting, the Project Owner may choose a cost-reimbursable commercial regime for the design consultant to drive better overall project outcomes. Table 10.2 provides an indicative breakdown of the salary multiplier, allocated into three limbs, being:

- Limb 1 – reimbursable costs (salary, employment costs and business unit operations costs)

- Limb 2 – corporate overhead and profit
- Limb 3 – pain/gain share.

Under this arrangement, the consultant is reimbursed their costs at the L1+L2 multipliers, with the L2 being at-risk depending on the L3 pain/gain share arrangements. Most consultants will recognise revenue at L1+50%L2 until there is certainty of the outturn performance and the likely L3.

Table 10.1 Indicative P&L statement

Item	*Measure*	*Recorded*
Gross Fee Income (aka revenue)	Fees invoiced on projects, less debts not received (or have not been received within 180 days)	Project
Direct Costs	Project expenses (less administration margin) such as sub-consultants, reimbursable expenses and other non-operating expenses	Project
Work in Progress (aka sales accrual)	Net Fee earned less Gross Fee invoiced and Direct Costs, subject to adjustment as described below	Project
Net Fee Income (aka Earned Income)	Chargeable time allocated to projects on the timesheets multiplied by the project rate adjusted by the work in progress	Project
Project Time Cost	Chargeable time allocated to projects on the timesheets multiplied by the project multiplier	Project
Project Gross Profit	Net Fee less Project Time Costs	Business Unit
Non Project Time Cost	Total employment costs less Project Time Cost • Selling • Administration • Training	Business Unit
Gross Profit	Project Gross Profit less Non Project Time	Business Unit
Overheads (aka General & Administrative expenses)	Operating costs, including: • Administrative salaries • Subscription and donations • Staff welfare • Recruitment and interview • Client entertainment • Public relations • Training • Property costs • Communications • Office supplies	Region

Table 10.1 (Continued)

Item	*Measure*	*Recorded*
	• Travel and subsistence • Motor vehicle expenses • Depreciation • Insurance • Audit fees • Legal and professional fees • Bad-debt provision • Computer costs • General costs • Sundry income • Bank charges • Asset disposals • Exchange differences • Recharges	
Operating Profit	Gross Profit less Overhead Costs	Region
Project Gross Margin	Project Gross Profit divided by the Net Fee	Project
Gross Margin	Gross Profit divided by the Net Fee	Business Unit
Operating Margin	Operating Profit divided by the Net Fee	Region

Table 10.2 Indicative multiplier breakdown

Item	*Description*	*Indicative Multiplier*
Salary	Salary divided by the standard working hours per week (e.g. 40 hours) for 52 weeks	1.00
Salary on-costs	Including superannuation, federal and state taxes and other staff entitlements (e.g. bonuses)	0.40
Business unit costs	Cost of operating the relevant business unit, including re-charges for services from corporate	1.35
Reimbursable cost multiplier	**Limb 1**	**2.75**
Corporate overhead	Costs upstream of the business unit	0.35
Profit		0.35
Overhead & Profit Margin	**Limb 2**	**0.70 or 25%**
Salary multiplier	**Limb 1+2**	**3.45**
Risk-reward	Percentage of margin at-risk	
Outturn multiplier	**Limb 1+2+/-3**	

Owners will often require a different multiplier for contract staff and overtime, given their different overhead rates. Also the multiplier for international staff is likely to be different, so it may be more convenient to treat international operations like sub-consultants as a reimbursable expense (with or without margin).

Key Performance Indicators

The typical commercial metrics or key performance indicators (KPI) for an operating business unit are outlined in Table 10.3.

The typical KPIs for a listed firm of most relevance to the shareholders compared to previous quarters or financial years are shown in Table 10.4. Analysts will review these metrics to assess the well-being of the firm and compare to the peer group. Honouring the forecasts provided to the market generates confidence in the strategy and performance and therefore greater investment. Greater investment means increased ability to fund strategic investments.

Table 10.3 Indicative commercial metrics for an operating business unit

KPI	*Measure*	*Recorded*
Utilisation (by hours)	Project Time divided by the standard working hours (or alternatively all hours worked)	Project
Utilisation (by cost)	Project Time Cost divided by the Total Employment Cost (Project Time Cost plus Non Project Time Cost)	Business unit
Salary multiplier	Net Fee divided by the Project Time Cost	Project
Productivity factor	Technical Utilisation multiplied by the Salary Multiplier	Business unit

Table 10.4 Indicative commercial metrics of a listed firm

KPI	*Measure*
Gross Revenue	Top line income
Project Gross Margin	Margin earned from project work done
Sales, General & Administration (SG&A)	Costs of running the business
Earnings Before Interest and Tax (EBIT)	Operating profit before interest and tax
Earnings Per Share (EPS)	Net profit per shares owned
Backlog	Work awarded, but not yet done
Operating Cash Flow	Cash in hand
Day of Sales Outstanding (DSO)	Days before the cash is received

10.2 Revenue recognition

During monthly operations reviews, financial auditing or during due diligence activities, revenue recognition of professional services firms can be somewhat problematic. This section aims to provide some principles for recognising probable revenue and known actual and accrued costs. For listed firms the biannual and quarterly reviews have become almost as rigorous as the annual audits. So the process has become expeditious monthly reviews, robust quarterly audits for efficient and expeditious annual audits and reporting.

Timesheet completion

The challenge for any professional services firm is to record the 'right' amount of chargeable time on the timesheet. For time and expense (T&E) projects this is relatively straightforward.

For lump-sum projects, the right amount of time is somewhat subjective and should be at the direction of the project manager. Lump-sum projects need to be completed within the pre-agreed timeframes and revenue recognised at the project multiplier.

The chargeable time determines the Earned Income, as identified in the chart of accounts above. It is therefore imperative that timesheets are completed and submitted by the end of each week. Many firms require that all worked hours are recorded. Others cap the time at the standard working hours, unless approved otherwise. Recognition of overtime is a complex issue.

On fast-tracked projects with a tight programme, sometimes overtime of selected personnel is warranted rather than deploying additional staff. Overtime needs to be managed and should be pre-approved by the line manager. Excessive overtime can lead to a loss of productivity and fatigue issues, particularly for staff that spend most of their day in front of a computer screen. Staff can be reimbursed for the overtime hours, or take it as time in lieu.

An indicative overtime framework is as follows:

- Staff should not exceed 12 hours per day (e.g. 8 hours regular, 4 hours overtime)
- Staff should not exceed 14 hours per day including travel from home to work and return
- Staff should not exceed 60 hours per week, without a rest day
- Overtime should be pre-approved by the line manager.

Earned income

For most professional services firms, income is earned when project time is entered on the timesheet. That is, fees generated from labour and direct costs are earned on an accrual basis, not a cash basis (when billed). Small practices often use cash accounting rather than accrual accounting. It is therefore imperative that timesheets are completed, submitted and approved at the end of each and every week. Income is evaluated as:

Earned income = project time x salary x project multiplier

As explained above, the key is to record the 'right' amount of project time on the timesheet. The other parameter is productivity. That is, not only capturing the right amount of project time, but also completing projects expeditiously, completing the project quicker and commencing the next project sooner, thereby generating more income in a financial year.

Project time not captured on the timesheet is effectively a write-off. Also, project time captured, but which cannot be invoiced or collected will be a write-off. Professional judgement is required to record the 'right' amount of time on the timesheet, which is outlined as follows.

Time & expense projects

All project time should be recorded on the timesheets, including project time in excess of the standard working hours, irrespective of whether overtime is paid or not.

Reimbursable project expenses include all the accrued direct costs in addition to normal operating overheads – such as printing, travel, accommodation, sub-consultants, specialist software – and should be claimed when the purchase order is placed and the costs incurred, not when the invoice from the supplier is received.

Lump sum projects

With well-planned and managed projects, project time should align with the budget estimates developed in the proposal. However, we know that with all projects, the actual progress is often different to that originally forecast. To align the actual expenditure with the planned, it is important that personnel understand their budget and manage accordingly.

Variations to the lump-sum fee

For approved variations, all project time should be recorded in the timesheet.

For unapproved variations, if there is an instruction to proceed and in our opinion the variation will most likely be approved in accordance with the contract, then all project time should be recorded, given we are recognising revenue at-risk and this should be approved by the line manager.

For unapproved variations, we should generally not commence services until there is an agreement. However, often we are required to continue the services under the contract, while the variation is agreed. If we are progressing the services at-risk then a separate activity code should be created on the project. If the variation is not approved due to concerns about the firm's performance, the project team should amend the work outside of the standard hours, or record time without recognising the income.

Work in progress adjustment

Work-in-progress (WIP) is the difference between the billing and the earned income and direct costs. Billings greater than the earned income generate negative WIP. This may reflect billings in advance, build-up of contingency and a potential write-up. Conversely, billing less than the earned income is a risk and may reflect a potential write-down. This should be assessed in each accounting period in the project performance report.

At the end of the accounting period, the WIP is re-evaluated based on the progress of the project compared with the budget. The WIP is monitored each and every month by the line manager. If the estimate at completion (EAC) varies from the budget, the earned income shall only be adjusted (write-up or write-down) at the approval of the line manager. This shall be recorded in the monthly review form. Material adjustments will be forwarded to the next line manager.

Progress is generally assessed as the percentage complete measured against either of the agreed fee, programme, deliverables issued or other milestones as stated in the contract. At the end of each accounting period, the project manager makes a declaration of the progress against budget and this is subject to review by their line manager. This assessment should not only consider the progress to date, but also the forecast fees required to complete the project.

Although there is generally no accounting standard for the valuation of WIP for professional services, the assessment may refer to AASB111 (Australian Accounting Standards Board, 2010) for construction projects or AS4817 (Standards Association of Australia, 2006) for progress assessment as guidance.

Based on generally accepted accounting principles, the adjustment of the WIP should be recognised in the accounts when the profit (or loss) has been incurred. However, professional judgement is required and this is subject to approval by the line manager and independent financial audit at year-end, which is explained as follows.

Prior to accounting for the write-down, the project manager should use every endeavour to re-engineer the project to complete within the EAC. If the EAC forecasts that the budget is to be exceeded, this write-down should be accounted for at that time.

Conversely, prior to accounting for the write-up, the project manager should consider potential risks that may still arise and may prudently declare to carry forward the potential write-up as contingency until completion and all debts receipted.

With regard to variations, the budget and the EAC should not be adjusted until the project manager has received written approval from the client, pursuant to the contract as outlined above.

Billings

Just as deliverables should be provided in the client's format, so should our invoices for expeditious payment and be issued on the date due, not at the end of our accounting period or calendar month. The covering tax invoice will show the amount claimed and the backing sheets should provide a breakdown of the progress claim to the detail sufficient for the client to process immediately. Extensive detail is not necessarily better.

As outlined in the CRM section in Chapter 5, the monthly progress claim should be used as a status report to discuss progress face-to-face with the client and seek feedback at that time.

Invoices should be prepared on an accrued basis of labour, direct costs and reimbursable expenses.

In submitting the proposal and agreeing to the contract, we need to be very mindful of billing for the work undertaken on an earned-value basis, rather than being constrained to milestone payments based on the specified deliverables. Better still, we should endeavour to bill against an agreed cash flow, with a mobilisation payment with a payment schedule to be cash positive over the life of the project. We are a consulting firm, not a bank or finance house.

Collections

Many businesses are profitable on paper, but are starved of cash to adequately run the business, with excessive interest charges and debt financing. Salaries and other costs are paid out each month and it is imperative that we receipt payments as soon as practically possible. Cash is the firm's oxygen.

The project manager is accountable for ensuring collections are received in a timely manner.

Unfortunately, clients are not always diligent in processing our invoices and this is reflected in our debtor days. Many clients need prompting to ensure the payment is made within the contracted timeframe. The commercial team can assist the project manager. The next invoice issued should record the payment history as well as the progress claim history and again late payments should be discussed at face-to-face meetings.

Sometimes payments are delayed due to either poor presentation of the invoice or perceived performance of the services. Again, this reinforces the need to personally present the invoice to the client each and every month and resolve any issues as soon as they arise.

Many countries now have a Security of Payments Act to stop clients from withholding cash without there being legal entitlement to set-off fees.

Poor working capital is often the consequence of a problem project as is lag data. This highlights that issues need to be addressed and escalated before generating significant WIP and debts outstanding. Operational managers should closely monitor their working capital days being the combination of WIP and Days of Sales Outstanding (DSO).

10.3 Fee estimation

A key commercial risk is completing the services within the estimated fee. A first principles bottom-up fee estimate is based on the following elements:

- Labour costs of the resource plan for each discipline at the project rates
- Cost of project management
- Contingent costs as a discipline (aka contingency)
- Project expenses, such as printing, travel and accommodation, software
- Sub-consultants, often with a margin
- Finance costs, such as escalation, expat on-costs, foreign exchange, cash flow and taxes.

The discipline leads will estimate their resourcing for their scope of services. Most bottom-up fee estimates will most likely include 'inherent' contingency. That is the natural conservatism of a first-principles estimate. 'Contingent costs' are the likely additional effort for known, known-unknown and unknown-unknown risks, as well as risks arising from the contract conditions.

For larger projects, a quantitative probabilistic risk assessment should be undertaken using a Monte Carlo simulation such as @Risk to evaluate the contingent costs (Palisade, 2017).

The project risk register should be based on ISO 31000:2009, Principles and Guidelines (International Standards Organisation, 2016) addressing functionality, cost, schedule, operational and reputational uncertainties. The risk management approach is to consider risk and opportunities on the following basis:

- Identification
- Assessment and analysis
- Mitigation or management measures
- Evaluation of likelihood and consequence (using a standard heat map)
- Treatment and controls.

This involves developing a risk register of identified likely risks, evaluating their impact, any mitigation strategies and then the likelihood (%) and consequence ($). The Monte Carlo probabilistic assessment considers the likelihood of all the risks occurring during the project. A P90 estimate would assume that most of the risks will be manifested. A P50 estimate may be more appropriate in a competitive marketplace and 'backing' the project management team to mitigate risks and harness opportunities.

The bottom-up estimate should then be 'benchmarked' against key metrics such as:

- Hours per deliverable per discipline (using a detailed WBS and deliverables list)
- Per cent of the construction costs (noting that previously published industry fee scales have been banned due to perceived anti-competitive practices).

Benchmarking similar and representative past projects will provide validation of the estimate or the likely Owner's budget.

10.4 Performance-based commercial regime

We briefly discussed pain/gain share arrangements above. This section provides further discussion on L3 and other performance-based commercial arrangements.

It is clear there are opportunities to develop better project-delivery methods in a collaborative relationship with the Project Owner. As a trusted project-delivery partner, we may consider that the commercial framework should help form the project culture and drive the right behaviours to achieve the project objectives, with interim and overall targets. As a trusted partner we may be prepared to put 'skin-in-the game' to share in the project success with win/win outcomes.

The framework in Table 10.5 provides an indicative commercial framework to progress developing this trusting relationship.

Table 10.5 Indicative performance-based commercial regime

Element	*Description*	*Type*	*Indicative commercials*
Direct costs	The size and mix of the delivery team will vary throughout the duration of the project, depending on the delivery and procurement strategies. In order to meet the varying demands of the project, there should be flexibility in the firm's resourcing. More when required and less when not required. The resource plan would be agreed with the client on a quarterly basis. It is suggested that the firm be compensated for the direct costs of deploying highly experienced and competent project personnel.	Reimbursable	Direct-cost rate for experience categories
Overheads	The overhead element should relate to whether personnel are working from the home office, the project office, or recruited specifically for the project working on the site.	Fixed	Either a fixed sum for a defined duration or $ per hour

Element	*Description*	*Type*	*Indicative commercials*
Expenses	Project expenses, disbursements and any specialist sub-consultants would be reimbursed.	Reimbursable	Reimbursed at-cost, plus a modest margin
Profit margin	Philosophically the firm recognises that incentivisation is a key motivator to achieving better than baseline outcomes and sharing in that success. Incentivisation is one of the levers for engaging the parties to a common purpose.	Fixed or variable	For poor project outcomes, the firm loses X% of the profit margin. For stellar project outcomes, the firm could earn Y% of the profit margin.

Performance regime

In agreeing any performance regime, some of the underlying principles that would form the basis of a partnering charter with the Project Owner would be as follows:

- A joint Governance Committee would be developed to provide governance and oversight of the relationship and performance of the work
- Develop a baseline plan with key indicative milestones and targets
- Jointly develop the performance regime measured against the project objectives
- Review the resourcing and performance on a quarterly basis, with a look-ahead to drive better than baseline performance
- Undertake all commercial arrangements and agreements in good faith
- Weight (aka gear) the KPI regime to emphasise the focus of the project team looking forward. Some examples are:
  - Safety and ethics
  - Environment and sustainability
  - Schedule
  - Budget
  - Quality, functionality and whole of life issues
  - Stakeholder relationships.

11 Contracts

This chapter describes the management of the contract agreements with the Project Owner, allied design partners, internal business units, sub-consultants and other sub-contractors and the complexity of appropriate risk allocations per Abrahamson's principles (Abrahamson, 2017). This is to balance the interests of the firm, client relationships and ease of contract administration. It also provides some guidance on successful contract negotiations for after and during the project to obtain the best outcomes for the firm and the requirement for robust and professional contract administration.

11.1 External contracting

As outlined in the Introduction there are often onerous conditions passed through from the Owner as a legacy of a buyer's market. Most consultants will have their own terms of engagement, however they are often rarely used and trading is most often on the Owner's terms, hopefully subject to negotiations in good faith. However, with many government clients it is not possible to amend the terms of the contract.

The general principles for key contractual terms for engagement of professional services are outlined below. The Owner's terms may not be fully covered by the proceeds of the firm's professional indemnity (PI) insurance or a project PI insurance policy representing significant 'balance sheet' risks. If we are required to accept risks beyond these standard 'terms of trade', they must be managed through the application of an approved management plan and priced into the risk contingency, either in the rates, as contingent hours or dollars in a lump-sum fee.

An indicative terms sheet for a professional services firm is provided in Table 11.1 and these terms are likely to be negotiated. The greater the scale and complexity of the project, the more difficult the negotiations are likely to be. The art of negotiation and the requirements for skilled negotiators is discussed later in the chapter.

Indicative liability regime

An indicative liability regime is outlined below. For most listed professional services firms, there must be an expressed limitation of liability. Drafting of today's

Table 11.1 Indicative terms sheet

Terms	*Commentary*
Warranties	The Consultant will not provide any warrant of a fitness for purpose nature. To the extent a direct warranty is required by us to the head client that warranty may not impose terms on us greater than what we agree with you.
Entitlements for Additional Fees	The Consultant is entitled to make claims for additional fees in the following circumstances: • Variations • Delay • Extension of Time • Rework or redesign (other than due to our default or error) • Suspension (other than due to our default) • Any other ground of claim under the D&C Contract.
Duty of Care	An acceptable duty of care would be as follows: The Consultant must, in the performance of the services, exercise all the due skill, care and diligence reasonably expected of a consultant familiar with local conditions, practices and requirements. The Consultant must carry out responsibilities in a thorough, skilful, and professional manner and in accordance with professional standards.
Responsibility for Information provided	The Consultant accepts that it has a duty to use reasonable endeavours to identify any erroneous information supplied to us but beyond a failure to meet that obligation, the Consultant will not accept responsibility for information or documents supplied to us.
Errors in Documents	The Consultant will only accept responsibility for errors in documents prepared by us, to the extent the error or omission is due to us. Our liability will be limited to the rectification of those documents.
Delay claims	The Consultant can claim delay for events beyond its reasonable control.
Novation	The Consultant may not be novated without consent.
Indemnities	The Consultant will not accept broad indemnities.
Limit of Liability	A Limit of Liability will be required in the following form: To the maximum extent permitted by law and notwithstanding any other provision of the Contract, the Consultant's liability to the client arising out of or in connection with this Contract (including the performance or non-performance of the Services) whether arising in contract, under statute, in tort (including negligence) or otherwise, is limited to aggregate to our Fee.
D&C Contract	Any other obligations from the D&C Contract that the client wishes to pass through must be specifically incorporated and agreed with us and would affect pricing. We will not accept incorporation of D&C terms into our agreement by reference.
Intellectual property	Grant of license for the purposes of the project only.

(Continued)

Table 11.1 (Continued)

Terms	*Commentary*
Insurance	The Consultant will provide evidence (as certificates of currency) of: • Professional Indemnity – $X million per claim and in the aggregate • Worker's Compensation • Public Liability – $Y million each and every claim.
Records	The Consultant requires the right at all times to retain one complete copy for its files and records for quality assurance reasons.
Other Unacceptable Clauses	• Releases as a precondition to payment or on completion • Set-off of claims or other amount which are not proven debts due at law • Recipient Created Tax Invoices • Time bars • Pure Economic Loss.

contracts, means that the cap on liability is often illusory or there are many carve-outs. All contractual liabilities should be included within the limitation.

Assuming the limitation of liability is set at the level of the PI insurance, Table 11.2 provides an indicative liability regime.

Most professional services firms will risk-adjust the liability regime based on the Owner's likelihood of claims. Most PI insurers or their brokers will advise that the most likely time of a PI claim is just before, or soon after practical completion. There is no doubt that the PI risks in the operational phase reduce once the asset is in the performance phase. Therein, why some project PI insurance policies do not cover the full term of the consultancy agreement.

Non-standard contract terms means non-fault-based indemnities, elevated standard of care, quantity growth, delay costs, liquidated damages, consequential losses. The standard exclusions to a liability cap would include personal injury, wilful default, gross negligence, criminal acts and IP infringement. Property damage arising from the performance of our services should be included in the limitation.

Commencing services without a signed contract

As contracting becomes more complex, the requirement to commence the services often occurs prior to agreeing and executing the contract. In these circumstances, a notice to proceed (NTP), or a letter of intent should be issued by the Owner. If none is forthcoming, we may consider providing a self-serving notification to proceed at-risk, with future ongoing negotiations to be undertaken in good faith and when the terms have been agreed to be applied retrospectively to the commencement date. All services would be performed in accordance with the latest version of the draft contract. Caution should be taken with this approach as many Owners won't pay invoices until there is a signed contract.

Table 11.2 Indicative liability regime

Client type	*Insured risks*	*Uninsured risks*
Government agency (with standard contract terms)	Proceeds of insurance up to $A	Unlimited
D&C contractors	Head contract claims: Proceeds of insurance up to the level of the head contract D&C claims: Proceeds of insurance up to the lesser of the Fee or $B	The lesser of Fee or $B Unlimited for standard exclusions
Private-sector clients (e.g. developer, industrial, energy, resources, utilities, operators, project managers)	Proceeds of insurance up to $B	The lesser of Fee or $B Unlimited for standard exclusions
Private-sector property owners	Proceeds of insurance up to $C	The lesser of Fee or $C Unlimited for standard exclusions

11.2 Design joint venture agreement

On larger design projects, design consultants will often partner and establish a design joint venture (DJV) to respond to the scale and complexity of the project. As these two firms are normally competitors, it is important to establish best-for-project (B4P) objectives and ensure the design team behave according to these agreed principles.

The design joint venture agreements (JVA) between the two firms will be based on the key principles outlined in Table 11.3.

11.3 Internal contracting

Contracting with internal operations or business units can often be problematic, given we often don't provide the rigour that we might do with the Owner, or sub-consultants. This section outlines some of the issues to be considered with contracting internally.

Purpose

When work-packaging to other business units, operations centres, areas or regions, the outcome is often disappointing. As an international consultant, work-packaging or work-sharing is one of the key aspects of managing workload. However, we often treat the internal client as second rate to our local domestic clients.

While it is important to further develop the collaborative model, there must be a clear line of responsibility and understanding of the risk sharing between the lead and the support functions. The purpose of this section is to provide

Table 11.3 Key DJV principles

Function	*Principle*	*Responsibility*	*Accountability*
Strategy & Governance	Two representatives from each DJV participant. Preferable to have one representative that owns and can commit the company-wide resources on a B4P basis, and the other with the KSEA in major projects. Provide clear demarcation between project and corporate management. One representative from each DJV participant to be appointed to the Design Leadership Team and for interaction with the Owner and other key stakeholders. Possibly consider appointment of an independent chair.	Compliance with the DJVA	Home office MD
Decision making	All decisions to be on a best-for-project (B4P) basis and be unanimous. Except for a defaulting party all liabilities are to be shared 50/50.	Compliance with the DJVA	Home office MD
D&C client	Map our Delivery Managers to the CJV structure and therefore a matrix structure. Preferable to have proportional representation from each of the DJV participants.	Timely delivery of design packages, using the DPR and the programme Compliance with the tendered design Design growth justification	Design Management
Technical Leads	Highly experienced technical discipline leads for compliance with the OPR and approval from internal and external reviewers. Consider mapping to the PS&TR. Selection to be made on a B4P basis.	Compliance with the OPR Resolution of technical issues with reviewers Safety in design	Design Management
Design Management	One senior representative from each DJV participant, with the requisite knowledge, skills, experience and attributes (KSEA) for a project of this scale and complexity. Provide allocation of roles and responsibilities in the DJVA.	Compliance with the CSA and the DJVA Responsibilities per the DoA	DJV Governance Committee

	Develop a delivery strategy (integrating with the Other Consultants) utilising allied consultants, secondments, and work packaging.		
Strategic Review Team	Two representatives from each DJV participants. Preferably a diversity of KSEA from technical to major project delivery. Consider utilising international specialists/experts. Consider also utilising strategic legal counsel for internal risk workshops, contract administration reviews and risk support.	Compliance with the Terms of Reference from the DJV Governance Committee	DJV Governance Committee
Internal Verification	One of the representatives to be responsible for the internal verification of the deliverables, with a dedicated independent team, prior to the three (3) external submissions.	Compliance with the OPR	DJV Governance Committee
Project Controls	One of the representatives to be responsible for progress assessment, payments claims, timely VN & EOT submissions, R&O management and CSA contract administration. Agree proven project control systems.	Progress assessment and reporting	Design Management
HR	One of the representatives to be responsible for the recruitment and on-boarding of the new project staff, sourced from the DJV participants, agencies, allied consultants.	Recruitment of all project staff T&Cs for all project agreements	Design Management
Internal Project Management	Each DJV representative appoints an experienced person responsible for the interface between the project and the home offices.	Responding to resource demands Submitting timely invoices to the DJV	Home offices

overarching guiding principles that should be reinforced in specific detail by the particular internal contracting agreement.

Clear line of responsibility

There should be clear distinction as to the Lead and Support office. The Lead office is the project manager of the internal contract agreement and the Support office the supplier of services, as specified on the BMS. That is, the Lead PM has responsibility for the project as a whole, including the sub-project elements.

The Lead office has the responsibility for preparing the internal contracting agreement and agreeing the allocation of the risks. The obligation of the Support office is to ensure the agreement reflects the understanding between the parties.

The Lead office will in turn have an agreement with the client, either another office of the firm or the Principal. The Lead office is responsible for ensuring this agreement is a reasonable form of engagement with the client and relevant elements are passed-through to the Support office.

Fee Type (in local currency)

In all instances, adequate scope definition, programme, fee and deliverables need to be clearly specified. Fee types include:

- T&E – time and expenses are provided to the designated activity code at an agreed multiplier or rate. The support office is not at-risk.
- Fixed Fee – the support office bears the risk/reward of over-runs and under-runs.
- Variations – the Lead office is responsible for directing the Support office to undertake variations.

In all instances, appropriate project management and direction by the Lead office is paramount.

Risks

The Lead office is responsible for agreeing the ownership of the risks. If not otherwise clearly specified and agreed, the Lead office will be responsible for all risks associated with the client agreement, including but not limited to scope definition, programme, quality, earned value, foreign exchange, variations, working capital, etc.

Working capital

A separate activity code(s) shall be established for the Support office. The Lead office shall allocate billings to the Support office on an earned value basis, in accordance with AS4817 (Standards Association of Australia, 2006) or similar and not accrue substantial positive WIP.

Ambiguity

Should there be any ambiguity or doubt when the internal contracting arrangements lead to two or more interpretations, it is usual to interpret it strictly against the Lead organisation under the doctrine of 'contra proferentum'. That is, interpretation is construed against the person making the statement; i.e. the benefit of any doubt in the meaning of the clause is to the benefit of the party that would be disadvantaged by it. That is, any ambiguity or doubt will be viewed against the Lead office for inadequate specification and agreement of the project requirements.

11.4 Contract management

As identified in the above sections, contracting with clients is becoming more problematic. Through effective negotiation, there is an opportunity to mitigate unreasonable risk allocation and obtain a more favourable outcome for the firm and determine the optimal fee. Too often we are polarised between the client's terms and the firm's contractual requirements.

This section outlines a practitioner's perspective, in lay terms, on some risk management strategies when seeking approval to accept client's terms that are less than favourable. This section aims to balance the client, commercial, contractual and legal requirements. Table 11.4 outlines some of the indicative contract management issues.

Table 11.4 Indicative contract management issues

Issue	*Response*
No go at the eleventh hour	Submitting losing proposals or proposals on which we can't come to terms with the client is a waste of effort and a poor return on investment. The initial go/no go process is an important 'gate' to ensure that we are able to submit a successful and binding offer. Medium- and high-risk projects will need a well-developed briefing paper with appropriate risk-management strategies for corporate approval. Although a conditional 'go' may be provided at the outset, it is conditional on negotiating reasonable terms with the client.
Proposal is an offer and may be binding	The proposal should include wording to the effect that we wish to discuss certain matters pertaining to the terms of agreement. Should the client accept our offer and therefore consider it to be binding, the offer is subject to satisfactory resolution by the parties on the outstanding matters, in good faith, prior to contract execution.
Proposed contract amendments is a non-conforming tender	Most mainstream competitors will seek amendments to the client's terms for medium- and high-risk projects. In the proposal, we should write in a conversational style that we would like to discuss certain contractual

(Continued)

Table 11.4 (Continued)

Issue	*Response*
	requirements, such as duty of care, fitness for purpose, limit of liability and third-party reliance. This should provide sufficient notification to then enter into negotiation should our proposal be of interest. Refer to the art of negotiation section for techniques for using perceived or actual leverage to achieve better outcomes. In negotiation, hold in-principle discussions first, discussing the risk exposure to both parties. Do not provide marked-up amendments in the first instance. Develop the heads of terms, providing suggested amendments or additional text.
Commencing services upon a letter of intent, prior to execution of the agreement	As time is often of the essence, clients may provide a notice to proceed or a letter of intent, subject to satisfactory resolution of outstanding items. We should confirm this notice as an instruction to commence the services, subject to satisfactory terms of engagement, in good faith, in accordance with our proposal. A defined process needs to be developed for then agreeing the contract terms in a timely manner. Revenue recognition will be at the judgement of the line manager.
Non-standard contracts	Non-standard client contracts will need to be reviewed by legal counsel.
Fitness for purpose	We are not insured for this open-ended warranty. We provide services that are suitable for the stated or specified purpose, as defined in our proposal. This highlights the need for a well-defined scope of services.
Ill-defined scope of services	It is important to define what we are doing (inclusions), as well as what we are not doing (exclusions). We should specify the outcomes to be delivered (deliverables), not just the process (methodology). For advisory assignments, where the deliverable is not validated from past projects, or codified, it is important to define the process in terms of inputs, process and outcomes, often referred to as the methodology. Like baking a cake, it is important to define the ingredients, the recipe and the type of cake expected. Whereas engineering projects are generally well defined and are modifications on previous practices, advisory projects are often characterised by an ill-defined process and outcomes. Advisory projects are normally secured by developing a unique and compelling methodology of the process to be undertaken to achieve the client's needs (or wants) as either implied or expressed in the request for proposal.
Elevated duty of care	We are insured for providing a standard of care similar to the industry standard. We cannot warrant the 'highest', 'best' or 'expert' standard.

Issue	*Response*
No limit of liability	The limit of liability, however so arising, should be limited to the level of our professional indemnity insurance. This highlights the need of our PI insurance to provide coverage for these risks.
Excessive professional indemnity insurance	PI insurance is a substantial cost. Similar to home and contents insurance, the lower the level of insurance coverage, the lower the excess and our risk profile, then our annual premiums will be lower. Our insurers will only provide certificates of currency, not policy details which are commercial in confidence.
Certainty of payment	Where possible, payments should be made on an 'earned value' basis, not milestones. Payments should be made in accordance with the agreed terms. If payments are not made in accordance with the contract, do not suspend services as this may incur delay claims. Rather seek legal and commercial advice to address the breach of contract. Some clients use liquidated damages as a penalty regime to improve our project-management practices. Liquidated damages are a commercial matter. By exception, we may agree to liquidated damages, but only if we are solely responsible for the delay, the client can demonstrate the loss arising and it is capped at a percentage of the fee.
Third-party reliance	We provide professional services to clients for the client's express use. If that client wishes to pass-through our deliverables to another party, that party shall rely on this information at their own risk. This should be clearly stated in our proposal and in our deliverables. Although we seek to be indemnified from any third-party claims, often clients are not willing to agree to this and we need well-developed risk management practices in place.
Client relationships	The conduct in the contractual negotiation reflects our other project management practices. We should always be professional and recognise that we can't always just accept client's terms as they stand. Sign up or walk away threats should be considered as veiled threats. To maintain the day-to-day project relationships, consider utilising the commercial manager or senior executive support. The next line manager should remain at 'arm's length' should wise-head dispute resolution be required. Before entering into a process on negotiation, ensure alignment with our client-relationship strategy.

11.5 The art of negotiation

As contracts become more complex, we need skilled negotiators to achieve the best possible outcome. The movie *The Bridge of Spies* (Spielberg, 2015) is a classic example of the art of negotiation. Expert negotiators need to balance competing interests of sales, delivery, legal and commercial operations.

Negotiation is a process, to achieve the best possible outcome (Roger Fisher, 1991). A negotiated outcome will always be a compromise, with a modest degree of discontent by both parties and we need to be mindful of what and when we concede, as this will provide an indication of our walk-way position. Refer to the BD and HR initiatives, to increase our awareness of the psychology of the relationship. The manner in which we undertake commercial negotiations reflects the professional nature of our services.

Negotiation is a two-way street

Negotiation is a process of give and take. When you give a concession to the client, always ask for something in return. This may be an additional fee. Core to our marketing initiatives, we most likely need to go first. To earn a trusting relationship with our preferred clients, we will go first – give a favour to get a favour. The client will visibly perceive that we are willing to be the first to make an investment in the relationship.

Remember you're the preferred service provider

If we are in the negotiation process, we will most likely be in a preferred position in the other eight elements of project management. While negotiating duty of care, fitness for purpose, limitation of liability, liquidated damages and the like, don't lose sight of the broader perspective and the other elements of the contract which are not in dispute. Be firm and don't concede too early, just for convenience – although don't protract the process.

Hold firm to your principles, utilise your leverage

Always take the moral high ground. As the services provider, we don't have the power in the relationship and are often the aggrieved party. Use corporate approval processes or the insurance broker as the silent third party. Strategise the leverage we have in the relationship; e.g. can the client afford to engage another party, can they walk away, have they conceded with other consultants?

Know when to seek assistance

Often a zippered approach works well, although most organisations will endeavour to resolve the matters within the project team. Also, wise-head dispute resolution will be through a greater spirit of compromise. Zipper the client, use in-house legal counsel, the operations manager and the next line manager are available to meet with their respective counterparts.

Don't be intimidated

Often contractors and developers like to intimidate consultants. They are masters at extracting 'value' from their subcontractors. Thumping the table and losing their cool are tactics often used to intimidate consultants. Be professional, not emotional.

Keep your ear to the ground

Two people should always attend these meetings. Always stay alert for clues that may tell where we are placed on the tender list and where they may be prepared to compromise. It's good practice to take notes of the meeting and a good sign if the client honours the actions and is punctual for the next meeting.

Respect the process

Stand in the client's shoes, be aware that the client representative needs to save face. To achieve a concession, he/she will need to substantiate this amendment. Provide them with the text to support their own approvals process. It is likely that any concessions will be subject to respective corporate approvals.

11.6 Contract administration

There is a falsehood that once the contract is signed it is then placed in the bottom drawer. It is the application of the contract that must be addressed by:

- Relationships and governance
- Requisite contract administration duties
- Management plans addressing attendant risks.

Managing our own contracts

Robust and professional contract administration is not inconsistent with being a client-centric firm. As explained in the foregoing sections, Owners often prepare onerous contract conditions for the downstream parties with onerous risk allocation. Our ability to claim Variation Notices (VN) and Extension of Time (EOT) will have strict time provisions (aka time bars). We must have sufficiently skilled and competent resources to address the contract administration functions. On larger projects a dedicated Commercial Manager is required to address the contract administration, project controls, financial management and reporting requirements.

Managing our client's contracts

When managing our client's contract as Superintendent, independent Project Management Consultant (PMC), or Independent Certifier it is important to clearly define in the contract documents whether the Superintendent is operating as an agent of the Owner, or as an independent certifier.

Construction lawyers such as MinterEllison (MinterEllison, 2016) will provide advice on the clear distinction between the two roles and attendant duties, risks and liabilities. An independent certifier is required to act impartially and without fear or favour, whereas an agent will be operating in the best interests of the Owner. The standard construction contracts will define the role of the contract administrator and most building and construction contractors will risk-adjust their tender price accordingly.

12 Risks and opportunities

> What gets us into trouble is not what we don't know. It's what we know for sure that just ain't so.
>
> (Mark Twain)

The foregoing chapters have outlined that an astute professional services firm needs to be prepared to take greater risk. This chapter outlines the issues with taking on more risk and how to be a prudent risk-taker.

The risk appetite should be applicable to the scale and competencies of the firm. For a large listed entity, shareholders, financial analysts, insurers and other stakeholders will require assurance of appropriate management, leadership, governance and oversight to ensure the precious reputation and sustainability of the firm is not placed at undue risk.

Continuity of knowledge is a key differentiator. However, past risks, incidents or events are not a predictor of the future. Project risks should be considered on a portfolio basis with client-relationship risks being assessed based on interpersonal, application of technical knowledge, and contractual (both implied and expressed).

Many major projects are failing to achieve their project objectives and heightened operational management and project leadership needs to be readily adapted to the heightened risk exposure. Mature firms should consider their operational management and project leadership as a competitive advantage and may be prepared to 'back themselves' and put margin at-risk, for better rewards.

12.1 Safety and ethics

Workplace health and safety

Our industry has developed in recent years in the pursuit of zero harm of our people and the firm's reputation. Most meetings are started with a safety conversation. In a professional services context, this is more 'well-being conversation' about themes such as:

- Slips, trips and falls
- Travel and journey management

- Mental health
- Fatigue management
- Respect in the workplace.

A safety conversation is most impactful when it is personal, heartfelt and authentic and is a key leadership quality. In our pursuit for zero harm, we now know that safety performance is directly linked to our behaviours, not simply applying processes. While we need to record the lag results of our safety incidents, the lead key behavioural indicators (KBIs) will drive the right behaviours that produce the desired outcomes. It is this 'cause and effect' model that creates the desired culture. This is described in more detail in the People and Culture chapter.

The following extract from the safety poem by Dom Merrell (2016) titled 'I could have saved a life that day' best articulates a behavioural approach to safety.

> If you see a risk that others take,
> That puts their health or life at stake.
> The question asked, or thing you say,
> Could help them live another day.
>
> If you see a risk and walk away,
> Then hope you never have to say,
> I could have saved a life that day,
> But I chose to look the other way.

Heinrich's triangle (Heinrich, 1931), often referred to as the 'safety triangle' showed that for every accident resulting in a fatality or major disabling injury, there are approximately 300 unsafe 'near-miss' incidents. The key message is that by reducing the number of 'near-misses' we will also reduce the chance of the fatality occurring, because we have lowered frequency of at-risk behaviours.

This awakening has changed the focus of operational managers and project leaders from not only measuring lag data (effects) but more importantly on the lead data (causes) many of which are behavioural.

In a professional services context this is most relevant when we are responsible for safety of a construction site and the safe operations of a complex industrial facility.

Safety in Design

Safety in Design (SID) is the integration of control measures early in the design process to eliminate or, if this is not reasonably practicable, minimise risks to health and safety throughout the life of the facility being designed. Many of the risks during the construction and operational phases can be designed out through careful and considered planning during the design phase.

Most jurisdictions have enshrined SID in their workplace health and safety (WHS) legislation. In most facilities, in particular industrial facilities, hazard analysis of the operational and maintained phase is undertaken. This will involve a HazID and HazOP risk analysis. HazID stands for hazard identification and is a general risk-analysis tool designed to alert management to threats and hazards as early in the process as possible. The classification made is done on the basis of probability and consequences. A HazID study provides a qualitative analysis of a workplace to determine its worker safety risk level. HazOP stands for hazard and operability study and is used to identify abnormalities in the working environment and identify the root causes of the abnormalities. It deals with comprehensive and complex workplace operations, which, if malfunctions were to occur, could lead to significant injury or loss of life.

Where it is not possible to design-out the risk, Project Owners may accept a safety analysis that provides the lowest possible risk outcome, known as So Far As Is Reasonably Practical (SFAIRP). In these circumstances, the onus is on the designer to assess the risks and build a safety argument around their design and make a recommendation to the Owner and/or Operator using a risk-based approach.

Ethics

Most senior operational managers will be incentivised to achieve the following targets or KPIs:

- Revenue, being sales and backlog
- Profit
- Cash
- Safety, and
- Ethics.

Although it is disappointing to not achieve the financial metrics, it is totally unacceptable to have a serious safety incident or ethical breach while on your watch. As outlined in our value statement in Chapter 2, we should be uncompromising in our compliance with safety and ethical standards. This relates to upholding our values as explained in Chapter 2. There is nothing more precious to a firm than its reputation. Most international firms will have stringent anti-bribery and corruption (ABC) practices, many of these requirements are enshrined in legalisation.

Most client contracts will have carve-outs to the liability cap for personal injury, breach of confidentiality and intellectual property rights, wilful misconduct and ethics breaches. Again this reinforces there is nothing more precious to the firm than its reputation.

12.2 Organisational risk management

Risk is defined by ISO 31000 (International Standards Organisation, 2016) as uncertainty, representing both threats and opportunities to achieving the business objectives. Risk impacts all aspects of the firm's operations, summarised in Table 12.1. This framework should be aligned to achieving the targets set in the firm's strategic plan.

Table 12.1 Indicative organisational risk matrix

Function	*Threats*	*Opportunities*
Business development	Market conditions or poor tendering yields insufficient sales and backlog to fund the capability or provides a delivery legacy	Increased tender success-rate and improved execution plans yield greater project success
Delivery	Margin erosion, claims and loss of client and partner relationships	Margin gain, repeat commissions, improved client satisfaction and reputation
People & knowledge	Poor selection of key project personnel; lack of team synergy; poor client and stakeholder relationships; and loss of IP	Improved project management capability and processes; development of technical excellence and high-performing teams; personal and professional development; incorporating lessons learned and innovation
Finance & operations	Drain on working capital	Support infrastructure drives improved efficiency and productivity with reduced overhead burden
Governance, risk and compliance	Breach of safety; ethics; and regulatory requirements; litigation and loss of reputation	Compliance with business management systems; values-based assurance; incorporating industry best practices

An indicative heat map of technical risks for a professional services firm is outlined in Table 12.2.

Research and experience suggests that the challenges to major projects are not simply technical knowledge, but the application of the knowledge in a project-team environment. The high exposure risks are often non-technical in nature and not often addressed in conventional risk-management practices. The human factors in developing a risk-based culture in the project management and governance functions are often the key ingredients to assuring project success. Claims are often for what we failed to do (negligence), rather than an error in what we did do. This is discussed further below.

12.3 Project risk management

> There are things we know we know, things we know we don't know, things we don't know we know and things we don't know we don't know.
>
> (Donald Rumsfeld)

Per ISO 31000 (International Standards Organisation, 2016) project risk management is based on the identification, assessment, management and control of

Table 12.2 Typical heat map for a professional services firm

	Consulting	*Design*	*Project management*	*Construction services*
Scope change	High	Med-High	Low	Low
Errors & omissions	High	High	Low	Low
Design growth	N/A	Very High	Med	High
Third party	High	Low	Med	Low
Certification	High	High	Med	Med
Client relationships	Med	Med	Med	Med
Sub-contracting	Low	Med	Low	High

Table 12.3 Typical qualitative heat map for a professional services firm

	IMPACT ->	*1 - Negligible*	*2 - Minor*	*3 - Moderate*	*4 - Major*	*5 - Severe*
Liklihood	1 - Rare	Insignificant	Low	Low	Medium	Medium
	2 - Unlikely	Low	Low	Medium	Medium	High
	3 - Possible	Low	Medium	Medium	High	High
	4 - Likely	Medium	Medium	High	High	Extreme
	5 - Almost certain	Medium	High	High	Extreme	Extreme

Risks and Opportunities (R&O). This is normally undertaken by developing a risk register identifying the causation of the risks, then the assessment of likelihood of it occurring and the consequence to the project. Risk identification of the known-known, known-unknown, unknown-unknown risks is often best undertaken in a workshop manner.

The risk assessments can be considered prior to, or after any mitigation measures are considered. The assessment can be undertaken on a qualitative basis. A common assessment is to use scale of one to five from very low to very high. A typical qualitative heat map is shown in Table 12.3.

For more complex projects, a quantitative assessment can be undertaken using a probabilistic analysis and a Monte Carlo evaluation.

Low-likelihood, high-consequence risks, known as Black Swan events (Taleb, 2007), should be considered outside the normal framework for risk assessment and allocation of risk contingencies.

For D&C tendering, it is now commonplace for the Tender Advice Notices (TAN) to also include a risk register, summarising the risk identified in the TANs. These risk registers are best undertaken against an elemental breakdown of the direct costs. If this is not available, then the next best method is to use the discipline breakdown from the Owner's project-requirements specification.

In summary for project risks:

- Risk is created through uncertainty
- We face technical and non-technical human-created uncertainty
- Astute risk management will have many benefits to the firm
- The global standard for risk management is ISO 31000
- The PMO should define the process for identification, assessment, management and mitigation of risks
- To be a strong and sustainable firm, all people should have heightened risk awareness as risk or initiative takers
- Independent Assurance provides a safety net in a riskier marketplace.

12.4 Independent peer review

> Questioning is the antidote to (implicit) assumptions that so often incubate mistakes
>
> (*Hon Charles Haddon-Cave*)

Independent peer review is an integral part of assurance of project success. An indicative Terms of Reference for an independent peer review of a D&C project is provided in Table 12.4. It could be readily adapted to other forms of project delivery.

Purpose

The purpose of an independent peer review at the early stages of the project is to ensure the project is set up for success and provide an independent review of the risks and opportunities to achieve that success.

Background

In design and construct projects, the 20 per cent completion stage is generally when the design is substantially complete, the project team is well established and the major procurement orders and subcontracts are known. An independent panel of experienced personnel can assist the project team in managing risks and harnessing opportunities, incorporating lessons learned and a fresh-eyes perspective. The review provides independent assurance through appropriate validation and verification that the project is being progressed as expected upon contract award, and in accordance with the management plans and business systems.

Conduct

The review panel will generally undertake a site visit and meet with key project personnel. The review would be unfettered and will vary from a high-level overview to a deep-dive depending on the level of confidence developed based on the enquiry. Supporting documentation would generally be provided in advance of

Table 12.4 Indicative agenda for independent peer review

ID	*Function*	*Elements*
1	Handover & mobilisation	Transition from tender and any risk-management requirements
2	Safety & environment	Leadership, culture, metrics and principal contractor duties
3	Objectives	Project objectives, targets, key performance indicators
4	Contract	Contractual obligations, milestones, liabilities (e.g. liquidated damages)
5	Management plans	Project execution plan, management plans and project control systems
6	Governance	Ethics and compliance, management board responsibilities and delegations of authority
7	Design (permanent works)	Design deliverables, milestones, approvals, safety in design, construction support
8	Design (temporary works)	Temporary works design
9	Construction	Construction methodologies, sequence of works, critical activities and productivity
10	Programme	Baseline programme, earned value assessment and progress curves
11	Cost control	Cost coding, earned value assessment, accrued costs, costs to complete and cash flow
12	Change control	Internal variations, budget transfers, client variations and extensions of time
13	Procurement	Key vendors, subcontractors, self-performed activities and forms of engagement
14	Quality and assurance	Compliance requirements, inspection and testing, completion documentation
15	Staff	Project organisation chart, key personnel, position descriptions and team performance
16	Labour	Workforce requirements, workplace relations, working hours, training
17	Plant, equipment and logistics	Major plant and equipment, stores and logistics
18	Reporting	Weekly updates and monthly progress reporting
19	Risks and opportunities	Risk identification, mitigation or management and opportunities for innovation
20	Warranties and insurance	Securities, warranties and insurance, defects close-out
21	Client and stakeholders	Relationships, communications and reputational issues

the review. The review is considered to be complimentary to the formal quality and financial audits, addressing all project-management functions.

The spirit of the review is to support the project team in achieving the project objectives and providing assurance to both the project and business stakeholders. If the project is being undertaken in joint venture, it is expected that the review panel will be sourced from the joint venture participants with complementary skills.

Indicative agenda

An indicative agenda is outlined in Table 12.4. It is expected that the chair of the review panel will determine the project-specific terms of reference.

Reporting

The review panel will prepare a concise report outlining observations with any areas of concern and suggested opportunities for improvement. A draft report would be issued to the project team for review and comment within seven days. The final version would then be distributed to line management and the project team. The chair of the review panel will follow up to ensure any actions arising have been closed out.

Conclusion

A copy of the reports will be included in the Project Management Office (PMO) as part of the ongoing lessons-learned programme. Depending on the scale and complexity of the project, additional reviews may be warranted at the 50 per cent and 80 per cent stages, along with independent reviews of the design phase.

12.5 Design growth

One of the biggest challenges in D&C and EPC contracting is quantity growth (aka design growth) arising from a preliminary tender design.

A D&C tender typically comprises a breakdown of costs as shown in Table 12.5. The aim is to minimise each of these elements, while satisfying the project and corporate objectives. For benchmarking purposes the costs are often compared to the Direct Costs (DC) on the assumption that the preliminary design has been accurately estimated.

This section addresses the risk of growth in these quantities from the tender phase to delivery phase, from a Consultant's perspective.

On many projects, the Owner's reference design is well-developed and the consultant is engaged to develop this into a design suitable for tendering by an experienced and competent D&C contractor. The tender design is a preliminary design based on limited survey, geotechnical investigations, equipment Vendor information and other design inputs. It is not a complete design and as such the D&C Contractor needs to consider known, known-unknown and unknown-unknown risks.

Who owns the contingency in the tender design? An overly conservative tender design will be non-competitive. The D&C team needs to determine the reliability of the tender design for:

Table 12.5 Indicative top sheet for a D&C tender

Element	*Tender Estimate (US$)*	*% of DC*	*% of CV*
Direct Costs (DC)	$305.00	100.0%	61.0%
Indirect Costs	$86.00	28.2%	17.2%
Design & CPS	$32.50	10.7%	6.5%
Risk	$17.50	5.7%	3.5%
Escalation	$9.00	3.0%	1.8%
Margin	$50.00	16.4%	10.0%
Contract Value (CV)	$500.00	163.9%	100.0%

- it to be relied upon to accurately estimate the Direct Cost of the Works for the tender submission and
- development of the tender design into a detailed design for fast-tracked procurement and construction purposes.

Social and economic infrastructure projects are highly competitive. Innovative or alternate designs are often a competitive advantage. However, these innovations are often less developed than developing the Owner's reference design, representing greater risk.

The design management methodology is the process whereby the Consultant works collaboratively with the D&C Contractor using every endeavour to minimise the Direct Cost of the Works.

Most D&C projects are fast-tracked. The Indirect Costs are generally dependent on the duration of the Works. The Consultant will assist the D&C Contractor to assess the constructability of the tender design and develop the most efficient sequence of the Works and how the design can influence the critical path.

It is to be acknowledged that the reliability of the tender design is highly dependent on the level of resourcing and effort required to provide tender advice suitable for pricing by experienced estimators. Some elements require small amounts of effort, whereas other elements require quite detailed analysis. The Consultant should recommend the level of detail required to develop the requisite level of confidence in the design deliverables, not simply design to a budget.

It is recommended that the Consultant support the D&C Contractor to establish the risks (and opportunities) arising from an incomplete preliminary design and develop plans for mitigation, management, or pricing of the risk. The Consultant should provide advice in relation to the preliminary design to assist the D&C Contractor in evaluating the R&O contingency. There should be a formal review milestone and the out-workings recorded as a Design Report.

The Consultant provides preliminary design information in order for the D&C Contractor to estimate the quantities to evaluate the Direct Cost of the Works. To cover the risk that the developed detailed design differs from the preliminary tender design, the D&C Contractor is required to assess a risk contingency and allow a contingent sum within the tender price. In the delivery phase, should this contingent sum be inadequate, this pure economic loss may not be covered under the Consultant's PI insurance.

In the tender-phase contract, the Consultant warrants that it has provided a professional duty in the development of the preliminary design. This duty is to provide due care, skill and diligence. It is acknowledged by the D&C Contractor that it is not a complete and detailed working design. The Consultant is required to procure a Professional Indemnity policy to indemnify the Consultant against the liability for breach of this professional duty. The PI policy covers the Consultant for breach of this professional duty and the direct loss arising from that breach.

As an incentive, the Consultant may consider putting some of the delivery fee at-risk, to share in the pain/gain associated with the outturn R&O contingency. This requires transparency in the Contractor's R&O allowance which is generally in the range of 1 per cent to 4 per cent of the contract value (or 4 per cent to 8 per cent of the direct costs).

On award of the tender, continuity of project knowledge from the tender phase to delivery is paramount. The Consultant should utilise the skilled and experienced professionals in the tender phase and commit those same key personnel to the delivery phase. Therein, the delivery phase becomes a validation of the tender design.

Experience suggests that working in an integrated team, co-located in a project office is the best way of developing a collaborative working environment between the Consultant and the D&C Contractor. Most Consultants have been involved in alliance contracting, transforming very functional project teams into high-performing united teams. However, with the primary focus and energy on minimising the outturn cost, experience shows that integrated teams will utilise greater amounts of effort in the design-development phase and this should be allowed for in the delivery fee.

No doubt design growth is a key challenge for the industry, therein why Owners transfer this risk to the non-owner participants. Design is a complex and iterative process. The Consultant should have the key personnel with the skill, experience and design processes to work with the D&C Contractor to address this risk in a planned and managed manner. There are many successful projects where the design management process has worked extremely well and there are valuable lessons learned when this has not.

12.6 Review and approval of design submissions

Another emerging key risk in D&C and EPC contracts is the delay from third-party approvers due to overly preferential and pedantic review. This is exacerbated if the reviewer is also the party that prepared to the Owner's reference design. The review is intended to confirm compliance with the Owner's Project Requirements (OPR). However, if the OPR is open to interpretation, this can delay approval or cause multiple iterations for design submissions with excessive comments and observations. Third-party reviewers include the Owner's technical advisor, Independent Reviewer, Verifier or Certifier, Government Agencies, Road/Rail Operators, Utility Services and Councils.

We are operating in more complex project delivery environments with multiple stakeholders and timely stakeholder approvals can be problematic delaying us from achieving a target date or milestone. The design review process is often specified in the upstream contracts. Often the D&C Contractor requires an internal submission prior to the external submission. We should therefore seek to limit the number of design packages and therefore the number of submissions. Splitting design packages creates significantly additional burden.

On the basis that we are warranting (aka assuring) our design submissions, any delay by a third party should be automatic grounds for an extension of time. We are incentivised to achieve the staged gate and claim payment.

The purpose of the Stage 1 preliminary or schematic design submission is to review and comment on the design for compliance with the OPR and for the Owner and stakeholders to provide comments for progression of the design. We should use every endeavour not to re-submit Stage 1 design submissions.

The purpose of the Stage 2 detailed design submission is to review and approve the design for compliance with the OPR or advise otherwise. The opportunity for preferential comments has passed. Preferential project enhancements should be the basis of a design change notice (DCN). This design submission is either compliant or not.

The purpose of Stage 3 final design submission is to finalise the design, incorporating any residual comments for construction purposes.

We generally have a 'right to proceed' to the next stage, provided we can demonstrate that we have addressed the progressive comments. We should refuse making re-submissions, unless there is a clear non-compliance with the OPR.

13 Knowledge management

> We should not look back unless it is to derive useful lessons from past errors, and for the purpose of profiting by dearly bought experience.
>
> (George Washington)

This chapter outlines the requirement for capturing project and organisational knowledge, for improved performance in the future. As a learning organisation, knowledge management is a core element of continuous improvement. The hard-earned lessons learned are from the firm, as well as industry peers. While we should learn from past experiences, with an uncertain future, their application is not necessarily a predictor of future success.

13.1 Lessons learned

Capturing lessons learned on a project is a process of understanding of:

- What went well?
- What didn't go well?
- What could be improved for the next time?

As projects are discrete linear activities, we as an industry are not effective at capturing lessons learned and imparting them onto the next project. With demobilisation of key personnel, lessons learned are often not well-captured. Also, with the hubris on the incoming team, they often consider they don't need to know the sins of others in the past.

Table 13.1 provides a 'dirty dozen' of lessons often not learned.

13.2 Learnings from across the industry

There is a growing body of knowledge that the larger the project, the more likely it will fail. Global spending on megaprojects is significant and they often fail to meet their objectives, in terms of schedule, cost and operational performance. By example, the Grattan Institute undertook a detailed study into the *Cost Overruns*

Table 13.1 Indicative summary of lessons learned

Issue	*Response*
1. Take the contract seriously	Utilise a contract execution process and enforce the management plan.
2. One-sided client terms	Responsibility for contract negotiation with the Project Governor, interacting with sales, legal, commercial and technical personnel.
3. Failure to claim VN and EOT	Enhance scope definition in the tender with an agreed change management process. Timely project reviews by the Project Governor.
4. Failure to agree IFC and RFI process	Management Plan to clearly demarcate completion of design documentation from technical advice during construction.
5. Failure to properly respond to allegations/claims	Enhance project reviews, empower and authorise the Project Governor to resolve issues before they become claims.
6. Failure to have appropriate dispute-resolution process	Utilise the Project Governor/Governance Committee as the first level of dispute escalation.
7. Failure to have proper document retention	Plan for the claim from the outset. Ensure contract administration diligence and rigour, with timely notifications. Ensure documentary evidence is robust.
8. Failure to project manage	Appoint the PM with the requisite skills, expertise, attributes, experience, with appropriate support from technical leads, project controls and Project Governor.
9. Agree to client demands without proper regard for consequences	Be professional, not servile.
10. Failure to appoint decision makers and keep informed	Appoint and empower the Project Governor. Develop a no-surprises culture.
11. Overly complex, not straightforward process	Use the PMO with proven business systems and processes, practice notes and guidelines.
12. Next one around the corner	Plan for the worst-case scenario from the outset (and it generally doesn't arise).

in Transport Infrastructure (Grattan Institute, Marion Terrill, 2016) over the past fifteen years in Australia. This research highlighted that Australian governments have spent AU$28 billion more on transport infrastructure than they told taxpayers they would. The cost overruns amounted to nearly 25 per cent of the total project budget. Or in simple terms, our society is receiving 7.5 rather than 10 assets promised.

Significant cost overruns are not limited to the Australian marketplace, or transportation projects. There is a significant body of knowledge that the larger and/or more complex the project the greater the likelihood of the project not achieving its objectives in schedule, cost or functionality. McKinsey & Company suggested there are three main reasons why infrastructure megaprojects go bad (McKinsey & Company, 2017b), namely:

- Over optimism and over complexity – lack of appreciation that the bigger the project, the more complex
- Poor execution – incomplete design, lack of clear scope, ill-advised shortcuts, errors in scheduling and risk assessments
- Weakness in organisational design and capabilities – from the Project Owner though the supply chain to the sub-contractors.

Other notable research on project success is from Bent Flyvbjerg (2006). The benefits only flow if the projects are properly completed. They rarely are. Cost overruns and benefit deficiencies of 50 per cent to 100 per cent are not uncommon.

From an Owner's perspective, projects appear to go wrong for two main reasons:

- optimism bias, and
- strategic misrepresentation.

Both lead to lower cost and higher benefit projections than are justified. This results in the least viable projects getting approved, because they look the best on paper.

It may be assumed that where optimism bias prevails, planners have an interest in rectifying its ill effects. Planners should also be subject to market and personal controls to prevent strategic misrepresentation of costs and benefits.

Optimism bias has four main causes, alone and in combination:

- Technical: imperfect information, bad techniques, poor data, honest mistakes, scope changes, management, poor risk analysis or other management issues
- Psychological: adopting an 'inside' rather than 'outside' view of the project. The inside view of the project is of its particularities, and problems. The outside view places the project within a broad reference class of historically similar projects and avoids planners thinking along the lines of '*this time is different*' without any real basis
- Economic: stakeholders have various interests in having projects approved, some interests conflict; this can give rise to what economists call principal-agent problems. If stakeholder A can make decisions that affect stakeholder B, and their interests conflict, winners and losers are inevitable
- Political–institutional: blindness to realities, the need/imperative to achieve funding or get work, 'monument complexes' or legacy building.

To avoid strategic misrepresentations, which mostly arise because of political or institutional pressures, then the major project requires:

- Transparency: make it clear that approving one project diminishes the pile available for other projects, independent peer reviews of forecasts require benchmarking, make estimates etc. public, there are professional, civil or even criminal sanctions for unethical estimates

- Market controls: the 'go-ahead' subject to private funding of 1/3 of projects without sovereign backing, full public financing to be avoided, planners should feel the pain of bad estimates, participation of risk capital should not mean government loses control.

When major and mega projects go wrong, everyone downstream gets caught. Professional services firms are often caught in the crossfire. There is an old adage that bad projects keep getting worse, and conversely good projects keep getting better. A 'no go' on a flawed major project may be the most strategic decision ever made by the firm.

Harnessing these learnings, Chapter 16 provides some insights on how to increase the likelihood of project success.

13.3 Case studies

Hypothetical case studies or war stories are an excellent method for sharing lessons learned and engaging a team in considering risks and opportunities for continual improvement. The following are some examples for each genre of professional services.

Case study no. 1 – consulting

The firm was engaged by a government agency to undertake a feasibility study for the remediation of a contaminated site in Melbourne. The site is subsequently developed by a developer and sold to residents. Some years later the residents complain of a pungent odour. Investigations demonstrate that additional remediation measures are required costing $20 million. The residents mount a class action against the government, the developer and the firm. The claim against the firm is made under the Competition and Consumer Act (CCA) that the feasibility report was deficient based on misleading and deceptive conduct. Under what circumstances would the firm be liable for these losses?

Case study no. 2 – design

The firm was engaged by a D&C contractor to prepare a tender design for the upgrade of the Pacific Highway in NSW. The D&C contractor's tender is the lowest and is awarded the contract. Towards the end of the contract, the D&C contractor claims against the firm that the as-built quantities are greater than tendered. Under what circumstances would the firm be liable for these losses?

Case study no. 3 – project management

The firm was engaged by a resources client to prepare a bankable feasibility study. The report is approved by the board and the project financiers and a subsequent EPCM contract is awarded to the firm for the delivery phase. The project does not progress as planned with multiple delays and cost over-runs. Many of the project contractors submit significant claims to the client for delays, additional scope and interference by the EPCM. The client claims against the firm for the project losses claiming that the firm failed to manage the project contractors. Under what circumstances would the firm be liable for these losses?

Case study no. 4 – construction services

The firm enters into a joint venture with a delivery partner to design and construct an upgrade to a wastewater treatment plant in Perth. The project is delivered late due to unseasonal weather and protests from aggrieved local stakeholders. The client imposes liquidated damages for failure to deliver by the date for completion. Under what circumstances would the firm be liable for these losses?

So looking back for the purpose of achieving better outcomes looking forward, the purpose of these case studies is not to find the answer, but rather engender a process of identification of what could go well (opportunities) and what could go badly (risks) on my project. This is best undertaken in a team environment in a workshop format.

14 People and culture

> We don't manage projects, we manage people to achieve the project objectives.
>
> (Tim Ellis)

This chapter describes the organisational and project issues with managing people and the requirements for developing an engaging and motivating culture. Professional services firms are people businesses and managing human beings (not human doings or robots) is complex and requires astute organisational management and project-leadership skills. It discusses the organisational arrangements, how to develop a sustainable business through accession and succession planning, how to solicit clarity of roles and responsibilities and how to create a culture of high performance.

14.1 Accession and succession planning

There are many examples of inspirational leadership and many when the manager has not engendered respect from their direct reports. It can be very disengaging working for someone who has no actual emotional intelligence or empathy. The old adage is that people leave their manager, not necessarily the firm.

It's not uncommon for firms to promote people into management because they're a star technical performer. However this does not mean they will be a good operations manager (Walker, 2017).

Promoting a good technical practitioner does not imply they will be a great manager. More often this is not the case. Good technical practitioners often focus on their own performance and their own personal targets and goals and that does not necessarily reflect their ability to manage and lead people.

Good technical practitioners can be intimidating as bosses, given they want to prove they are the 'smartest person in the room'. Occasionally they can come across as bullies. They tend to hold themselves to a high standard. And if they are the best performers in an organisation, more often than not others won't be able to reach those standards and they're not forgiving if they don't.

A good operations manager exhibits the following attributes:

- Their ability to provide direction, and that's either through a vision, strategy, objectives, or goals
- Their ability to motivate and create an engagement and commitment within their team
- Their ability to develop their staff though motivation, direction and coaching.

Coaching and development is something many professional services firms' companies don't do well. This is not training and instructing. Smart and clever technical people require evidence to learn. This evidence is provided through watching the person perform on the job to see that they have the capability to do that role or the capability to improve. This is often the missing ingredient in training courses.

As explained earlier in the chapter on Business Development, emotional intelligence is the ability to read people and personalities and also to know that at different times and circumstances the manager may have to act differently. The operational manager of the future will be an adaptive manager.

In terms of performance management, if you have a staff member who is highly capable but who is not performing well, it could be because they haven't been given good direction, or they may be demotivated or disengaged and it may not have anything to do with work. It could be something that's happening externally.

They might need a different type of support to get them performing to their potential again and it takes a good manager to be able to talk to their team and to discover and understand which of the components they need to work on to supply that support and to get that person functioning to their best ability again.

So some people have the right DNA to be operational managers. For the highly technical practitioners, there must be a career pathway for them as well. And this should be the primary pathways if we aspire to have a professional service firm, known for its technical excellence. Accession planning should be based on enhancing the reputation of the individual and the firms and providing technical leadership across the organisation. Empowering this horizontal slice through the organization will promote technical excellence and lead to improved technical quality and fewer errors and omissions.

For the reasons provided above, highly competent technical practitioners often make greater coaches and mentors on technical matters and this should be exploited and part of a structured succession planning process.

14.2 Responsibilities and accountabilities

Whether an organisation or project organogram, it is imperative that the roles and responsibilities of the key personnel are defined in the management plans and effectively communicated. The management plans are the music that keeps the orchestra in harmony. The organogram is not a wiring diagram of command and control, but a network of relationships.

To understand the inter-relationship and reliance of other team members a RACI matrix (Project Smart, 2017) is an effective tool. Responsibility (R) is allocated to the person responsible for the task or activity, accountable (A) to a nominated person with the designated authority for approval. The responsible person will need to communicate or consult (C) with other team members, whereas some personnel will only need to be kept informed (I). A RACI matrix can be a powerful tool for building collaboration and communicating across the boundaries and interfaces that often impede organisational or project performance.

The word Responsibility has two parts: 'response' and 'ability'. When you are responsible you totally recognise that you and only you are in the most powerful position to respond to whatever is going on in your life. Being responsible means we do not look outside of ourselves for a solution to whatever is happening in our life. This thinking takes a lot of personal work and an ability to accept that there is always something that we can do. Consider that when you make the choices you make today they have a rippling effect. When we are able to be present to and live fully with this, we become free. We experience a feeling of peace and harmony where we trust our own judgement and accept the consequences of our own decisions. The more responsibility, the faster you will correct the mistakes and poor judgements you make.

Being Accountable means we are able to hold others to the goals and results that they promised. This score is paramount in any team leader, manager or executive leadership position. A leader who cannot hold others to the results they have promised will affect the bottom line of the business. If you are not in leadership and work within a team, accountability is important, especially if you are required to produce results as a team. Your ability to notice and not 'walk past' unmet deadlines or promises provides the peer pressure to motivate the team. As the Irish statesman Edmund Burke stated, 'The only thing necessary for the triumph of evil is for good men to do nothing'. Being accountable means not doing nothing. As explained in Chapter 11, claims against professional services firms often arise for what we fail to do rather than what we did do in error.

Line management positions (square boxes) are accountable for measurable results. That is, they are readily 'accounted' for each and every month by performance against a target. The position descriptions are therefore quite prescriptive.

In a matrix organisation, off-line positions (circles) are influencing roles that have qualitative responsibilities. That is, responsibilities are performance-based obligations that are measured against achieving either project or business objectives. For off-line management positions, this would be achieving the project or business objectives. Position descriptions should be more concise, based around qualitative objectives than quantitative metrics. It is more about the person than the position description and therefore management is more an art than a science. As identified in Chapter 9, therein why many poor operational managers resort to management by metrics rather than management by objectives, given their poor people-management skills.

Table 14.1 Indicative governance attributes

Prescriptive	Performance-based
Management	Leadership
KPI	Objectives
Directive	Influencing
Push	Pull
Requirements	Requests
Controls	Empowerment

Off-line management positions are very important within a multi-dimensional business for cutting across the silos. They are the 'boundary-riders' that glue together the silos and cross-fertilise the business for the benefit of the client or the organisation, with a best-for-project outlook. Any person in an off-line position must have strong influencing and inter-personal skills to make a difference. They need to have strong emotional intelligence to build trust and rapport with internal and external stakeholders.

In terms of holding key personnel accountable, the distinctions between prescriptive and performance requirements are outlined in Table 14.1.

14.3 Creating a culture of project high-performance

Our sense of purpose

A culture of high-performance in a project environment arises when there is close collaboration between the project participants.

Project teams are often sourced with highly intellectual and technically-qualified personnel required to address the technical aspects. These staff are generally highly left-brain biased and can often be low in right-brain emotional intelligence. Similar to Chapter 5 on Business Development, project success can be defined as:

$$\text{Project success} = (IQ + TQ) \times (EQ + SQ)$$

That is, harnessing the intellectual qualities (IQ) and technical qualities (TQ) of the team members with emotional qualities (EQ) and spiritual qualities (SQ) of the collective (Newman, 2009). Emotional qualities include assertiveness – providing clear and concise decision making; optimism – providing a positive outlook; self-awareness of the influence on others; self-confidence – exuding high levels of self-regard and self-esteem; interpersonal relationships providing strong working relationships with all stakeholders; empathy – we can stand in others' shoes and understand their perspective; energy – creating a passion and energy that inspires the project team; and resilience – having the mental toughness to overcome challenges and risks.

In a project context, spiritual and social qualities are having a common sense of purpose, shared values and the focus of the collective well-being and not the individual. How can we be happy in our career, our personal and professional relationships, live our lives with integrity consistent with our values and have an enduring source of happiness? Fredrick Herzberg (1959) suggested that the most powerful motivator is not money. It's the opportunity to learn, grow in responsibilities, contribute and be recognised and valued. As explained earlier, that is why mentoring and coaching the emerging professional is noblest of all our duties.

If we are not guided by a clear sense of purpose, we are likely to waste our time and energy on matters that aren't really important to us. Responding to the challenges identified in the Introduction to play a vital role in the social, environmental and economic well-being of our economies provides that sense of purpose. We are instrumental in the development of the vital role infrastructure plays in the liveability of our cities and most urban conurbations.

Our way of being

Understanding the neuroscience of how the brain works will improve communication and management and our ability to influence and motivate others.

There has been much research and science, as well as ancient knowledge and practices of martial arts Eastern philosophy, that help us understand how the brain helps or hinders effectiveness, relationships and change, how emotions influence thinking, behaviour, performance and decision-making. These understandings should deliver better engagement and performance.

Projects are essentially the melding of people and processes to achieve the desired product. The people component is not only having the right people with the requisite knowledge, skills, experience and attributes (KSEA), but also ensuring that they interact effectively. Human interaction is a core process and is the key ingredient to achieving the project targets. Human interaction occurs through conversation and relationships and so a project can be considered as a network of conversations and relationships and they are the heartbeat or lifeblood. The quality of the conversations determines what actions get done and how well they are performed.

Ontology is the philosophical study of the nature of being, becoming, existence or reality as well as the basic categories of being and their relations (Siler, 2016).

It helps understand our 'way of being', why we behave the way we do and why others behave the way they do. It has three core elements:

- Language, generates reality through spoken words and our inner thoughts
- Emotions and moods, shape our perceptions, behaviours and our energy
- Body, is the embodiment of who we think we are and what we think is possible.

The verbal language we use, the emotion we exhibit and our body language are all signs of our way of being. A highly astute and emotionally intelligent project leader will be adaptive to their environment for the particular circumstance. As Charles Darwin stated 'It is not the strongest of the species that survive, nor the most intelligent, but the one most responsive to change'.

Mindfulness is the psychological process of bringing our attention to experiences occurring in the present moment. Operational managers with high EQ are able to 'sense' their influence and impact on the people around them. Like most leadership qualities, mindfulness is a learned quality. Mindfulness is also highly relevant to safety learnings. Taking a step back and assessing the risks is a form of mindfulness. These behaviours can be used for all risk assessments. What could go wrong and what could go well?

Impediments to project success

Table 14.2 provides twenty examples of where our unconscious biases can impact our decision making (Business Insider, 2016).

14.4 Creating a culture of organisational high productivity

The following section presents some strategies on how operational managers can begin to take a more active approach in creating a culture of high productivity to achieve their targets (Grech, 2005).

To create this culture all people need to be engaged in the firm. Creating an engaged workforce requires regular attention and maintenance to the relationship between manager and staff. Central to many of the strategies below is the need for operational managers to maintain clear, open communication channels with every person, so there are shared goals and expectations. These are some suggestions:

Communicate the productivity culture

Schedule regular divisional meetings that outline the importance of lifting productivity and meeting agreed targets. Explain that it is fine to work in a fun and enjoyable environment; however, this is done on the basis that we work hard and deliver.

Increase effectiveness of work time

Establish boundaries to reduce interruptions and elevate productivity. People need to work 'concentratedly' without unnecessary distractions. Talk to those staff who continually interrupt others or are seen chatting, to remind them that they need to have respect for other people's time and need to focus.

Communicate profitability targets

Explain to staff the performance metrics in simple terms and what it means to them. Bring along the commercial team to help staff understand and show staff

Table 14.2 Cognitive or unconscious biases

1. Anchoring Bias People are overly reliant on the first piece of information they hear	2. Availability Heuristic People overestimate the importance of the information that is available to them	3. Group Think The probability of one person adopting a belief increase based on the number of people who hold the belief	4. Blind Spot Bias Failing to recognise your own cognitive biases is a bias in itself
5. Choice Supportive Bias When you choose something, you tend to feel positive about it, even if the choice has flaws	6. Clustering Illusion Bias This is the tendency to see patterns in random events, like gambling	7. Confirmation Bias We tend to listen only to information that confirms our preconceptions	8. Conservatism Bias Where people favour prior evidence over new evidence or information that has emerged
9. Information Bias The tendency to seek information when it does not affect the action	10. Ostrich Effect The decision to ignore negative information by burying one's head in the sand	11. Outcome Bias Judging a decision based on the outcome, rather than how exactly the decision was made	12. Over Confidence Bias Some of us are too confident about our abilities and this causes us to take greater risks
13. Placebo Effect When simply believing that something will have a certain effect on you causes it to have that effect	14. Pro innovation Bias When a proponent of an innovation tends to overvalue its usefulness and under-value its limitations	15. Recency Bias The tendency to weight the latest information more heavily than the older data	16. Salience Bias Our tendency to focus on the most easily recognisable features of a person or concept
17. Selective Bias Allowing our expectations to influence how we perceive the world	18. Stereotyping Expecting a person or group to have certain qualities without having real information	19. Survivor Bias An error that comes from focussing on surviving examples causing us to misjudge a situation	20. Zero Risk Bias We thrive on certainty and not taking a risk is a risk

the monthly results. Identify and explain what the targets are for the week and the month. Also, sit down with project managers to monitor profitability of projects.

Demonstrate behaviours

Be sure to demonstrate the behaviour you want to see occurring in your business; i.e. hardworking, focused, committed, etc. Remember your leadership influences others in the business. In your management meetings, reach consensus between key personnel on what are the agreed behaviours that you want to exhibit.

Hands-on management

Take a more hands-on management approach with staff during critical periods. Walk the floor at least every couple of days (schedule time in the diary to do this) and ask questions to staff about their projects, how they are tracking and what they are working on, etc. Let them know that you are interested and are monitoring what they are doing with their time. Ensure the right number of staff, with the right skills are working on relevant projects.

Reward and recognition

Reward and recognise commitment and the right behaviours exhibited by staff and let others know about it. Rewards can be praise and acknowledgement in front of their peers and possibly the client and to senior management, reward in-kind tailored to the employee's interest, or a monetary sum. This is only to be given in circumstances where there is outstanding performance.

Meetings management

Consider the value of who is included in meetings and which staff need to be at meetings. Decisions in meetings can sometimes be made by just visiting someone's desk for a two-minute conversation. Encourage others to be wise and efficient with the time of others and themselves. Stand-up meetings, or a walk around the block could be more effective.

Efficiency

Identify non-productive and under-utilised staff with the team leaders and put together a plan to lift performance over the next few months. Set targets for utilisation and consider talking to other teams that are in need of staff.

14.5 Empowerment versus controls

With an innovative and entrepreneurial spirit, the firm needs to be nimble and agile regardless of whether the job is small or large, close to home or abroad, and this is what would make us special to our clients and differentiates us from our competitors.

If we want greater empowerment of our key personnel and to expand the 'sieve' of risk controls and authority limits with the controls being less prescriptive, is it appropriate to consider what these key personnel should do when uncertain about how to interpret the guidelines? We don't want to 'throw the baby out with the bathwater' as we refine our business processes.

These decisions are underpinned by our risk culture. The desired culture is defined by the behaviours of our key and influential personnel, as well as our business processes.

In a negative sense, if someone worked the system, played the game or buried issues, then we would consider that behaviour as a breach of our integrity (aka trustworthiness) value and we should be able to hold personnel accountable for such actions or inactions through their employment contracts, in a 'just culture'. In a positive sense, we should reward and reinforce behaviours that align with our values such as early notice, no surprises, open communication, collaboration and escalation when appropriate.

So how do we get the balance right between empowerment and controls and how do we trust our people to do the right thing during times of uncertainty?

The following suggests there are three underlying principles to getting this balance right.

Risk-based culture

The first principle is a depiction of what a mature professional services firm would expect. A mature risk culture is present when personnel at all levels routinely anticipate risks and advise/report any issues of concern to the person they report to. For project work, that would ultimately be the Proposal Manager or Project Manager, or the Project Governor.

This is depicted in Table 14.3 (Deloitte, 2016).

Trusting relationships

Negligent, reckless and improper behaviours and complacency can materially undermine our performance leading to margin erosion, negligence claims, and loss of key clients. Trust between us and our clients and personnel within the firm is earned, based on four key attributes. Adapting David Maister's (2001) trust equation:

$$T = C \times R \times Q / SI$$

Credibility (C) is based on experience and competency of the key personnel. Reliability (R) is based on fulfilling commitments and being accountable for outcomes. Quality (Q) is based on the quality and depth of the relationships, expressed verbally, non-verbally and in the written form in our deliverables.

Table 14.3 Instinctive versus learned behaviours

	Detached	*Involved*	*Proactive*
Robust	DISMISSIVE 'There are so many rules, it is hard to get anything done in the time available unless you cut corners'	CONTROLLED 'I follow the rules and procedures that are laid down, even though they can be a bit of a straightjacket at times'	ANTICIPATORY 'I have a strong risk platform for my work and am stimulated to think about enhancements'
Adequate	APATHETIC 'There is enough leeway in the risk guidance that I can do my own thing when it suits'	COMPLIANT 'I follow what requirements exist, largely to avoid punishment for breaking them'	COMMITTED 'I seek to make good risk decisions and look out for others but gaps in our framework give me concern'
Limited	IGNORANT 'The company is just interested in getting the job done with minimal bureaucracy which suits me fine'	INQUIRING 'Guidance is lacking so I make judgements about what is best for me and what makes sense'	INHIBITED 'I make every effort to anticipate risk, but would appreciate more support from the firm, peers and PIC.

However, this trust can be undermined by Self-interest (SI), contrary to a best-for-project, or best-for-client portfolio approach.

As we empower our staff, it should be clear what constitutes a breach of trust, in a 'just culture'. Or in a positive sense an understanding of the 'bond' of trust (like an implied agreement). There is a balance between a 'no blame' culture, which encourages staff to report issues without fear of recrimination, and a culture of accountability. Accountability should not be interpreted as meaning someone to blame. A 'just culture' encourages open communication to promote and resolve issues expeditiously, understand lessons learned, improvement of the business processes and professional development.

Delegation is not an abrogation of responsibility, but with appropriate oversight, guidance, supervision and support, depending on the trust relationship between the delegator and delegate. For our projects or portfolio of projects, there should continue to be appropriate levels of governance, oversight and supervision by the person ultimately responsible, being the Project Governor.

Values-based assurance

The controls prescribed in the business processes need to be tailored to the perceived and actual risks and reviewed periodically. With a mature risk culture, risks are not transferred or allocated, on an 'all-care no-responsibility' basis. As explained earlier, when staff feel empowered and they are delegated responsibilities based on trust, it will be in an environment with appropriate oversight, akin to parental guidance or pastoral care. In this context, care means we are committed to the well-being of our projects and achieving the target project objectives.

Management, shareholders and clients expect that issues will be addressed expeditiously with assurance that risks will not be overlooked, we are not 'blind-sided', or they 'fall-through-the-cracks'. This assurance is provided by independent review, using a risk-based approach, based on the perceived and actual risks. The reviews can vary from a 'light-touch' to a 'deep dive' depending on the risk exposure.

The obligation of empowered personnel is to openly communicate during times of uncertainty and welcome independent assurance, being proud, passionate and anticipatory in the management of their risks. This culture of assurance is aligned with our corporate values, and therefore is referred to as values-based assurance.

In summary, this section demonstrates that appropriate balance between empowerment and controls is determined by values-based assurance, through:

- Empowerment of our personnel
- Developing trusting relationships
- Application of independent assurance.

Firms with a mature risk-based culture will outperform their peers. Risk is inherent in every decision we make. A risk-ready anticipatory business will outperform a risk-averse or risk-naive business. That does not imply a cavalier approach to risks, but an approach based on sound human behaviour and business-control principles.

It could be argued that risk warps the objectives in the strategic plan. Embracing a risk-based culture creates value by connecting risks to rewards. The professional services firm of the future will use risk management to fuel better performance.

14.6 The recruitment experience

As explained earlier, there is and will be a skills shortage in the infrastructure sector. The war-for-talent is expected to worsen as more professionals are exiting the industry than entering. This is particularly paramount in a highly competitive labour market. The purpose of this section is to provide strategic direction to increase our chance of success in landing strategic recruits.

It suggests that if undertaken properly, this process will create an enduring and loyal relationship. That is, alignment between the firm as the employer of choice and the candidate as the employee of choice. Similar to client engagement, the recruitment process should be a sensational experience.

Our growth as presented in the foregoing strategy map format, is predicated on organic strategic recruitment and M&A. Additionally, part of the recruitment process is also to ensure there is a cultural alignment between the firm and the candidate.

As with having a focus on delivery of a break-through performance for our clients, we need to ensure we give our candidates a sensational recruitment experience with the firm. The experience should be carefully planned with an eye to developing a long-term and enduring relationship.

The fishing analogy

The recruitment process is analogous to 'fishing for talent'(Grech, 2006). With a limited talent pool, effective recruitment is the difference between luring and landing the species rather than casting out berley in the hope that you get a bite. The talent pool is a living environment and it changes daily. There are different generational needs, people will change jobs more frequently, personal circumstances are varied, people arrive from interstate or overseas and people retire.

Simply sticking a line and bait into the water may be a good way to have a relaxing afternoon, although it's a lousy way to catch fish. Therefore, it is important to keep in touch with movements in the employment market. By being aware of changes, such as senior appointments, acquisitions, it allows you to approach recruiting in a strategic way, and may give you a better indication of lucrative 'fishing holes'. Once the word gets out about good fishing holes, other fishermen move in and it's time to find another hole. When a huge number of fishermen fish in the same spot repeatedly, the 'hole' gets fished out. Consequently, as the competition increases, the quality of the lure and use of advanced techniques make the difference.

Keep in mind that sometimes the fish are biting, sometimes they're not. Professional fishermen use expert techniques and fish at different times and different locations from amateur fishermen.

Central to the analogy is that the bait is a proposition to the candidate. Effective attraction is based on our invitation to participate in an exciting journey with the firm. How do we create an enduring and trusting relationship? Again referring to David Maister, to earn a relationship, we must go first: give a favour to get a favour (Maister, 2001). The one you are trying to influence must visibly perceive that you are willing to be the first to make an investment in the relationship.

Trust-based relationships are based on robust dialogue, ownership and accountability and a focus on results that define the essence of the relationship.

Suggestions to effective fishing for talent include:

- Consider where current high-performing employees have come from and why they have been most successful with the company. This may give you a better indication of quality fishing holes.
- Assist in pre-recruitment activities to build the firm's visibility in the employment market. This helps generate awareness around the opportunities we offer and ultimately can lure interest in our firm.
- Stay up-to-date with changes in the employment and acquisition market, by reading websites, attending seminars and reading industry magazines.
- Advising key recruitment agents of the position and its details. The more knowledge they have the more effectively they will choose the right bait.
- Ensure it is listed on our Careers website and Internal website.
- Ensure vacancies are communicated to all staff. Word-of-mouth advertising can create more 'fishermen'.

Fishing is a sport of patience and tenacity. Sometimes you can catch the perfect fish straight away, but most times it requires great thought and consistent energy. With the current labour shortage, don't just wait for the fish to come to you. Ensure you partner with HR to review your fishing techniques and plan your fishing trip.

Engagement

Once the fish has taken the bait, reeling the fish in is an art in itself. Traditionally candidates have been viewed from a perspective of 'Why should we hire you? and 'What can you do for us?' However, in today's buoyant market it's a necessity to market to the prospective employee the career, employer brand, position description and its benefits. We need to be cognisant of their needs and wants.

This is often left too late in the process and the candidate may lose interest and decide not to progress. It's important to remember that both parties to the recruitment process are buyers and sellers. Too often, companies treat candidates as mere job-seekers or job-shoppers, erroneously assuming that there is no competition for a given candidate's attention. The line should not be too taut and break, nor be too slack such that the fish slips away.

Strategic process

Most employers will fish in an inconsistent and subjective way. The fishermen involved in the recruitment process need to decide on the strategy employed to reel the fish in. This may include the number and type of meetings, questions asked and even who is most appropriate to conduct interviews. Significant thought should be given to creating a compelling value proposition to the prospective candidate. The central question that must be answered is: 'Why would a candidate who is happy, successful and fairly compensated leave a quality situation and come to work for us?'

The key to answering the above question is two-sided. To hook the fish initially, as fishermen we need to put a certain amount of energy into selling the role and company. Once this is done, then we need to structure the engagement process to effectively determine what would be the key drivers for candidates to move employers. Identifying these drivers through effective questions and speaking to referees, will allow you to shape an offer that meets the candidate's expectations. However, it is highly important that the candidate's expectations are in line with the job role and business requirements.

Candidates who are hired should have at least one to two meetings with the prospective employer, where the candidate does about 80 per cent of the talking. Reeling the fish with a plan will give the fishermen an opportunity to understand the capabilities and shortcomings of the candidate and assist in developing a compelling job proposal.

A plan should be in place outlining how each meeting will add to an in-depth understanding of the candidate's fit to the role and company.

The first impression

You will never get a second chance to make an everlasting first impression. Whenever possible, hold your initial meetings at a neutral location away from your normal workplace. Serious candidates will want to see the company in action, but the first session is meant to whet your appetites for further discussions, not answer every question that you or the candidate may have. By putting distance between your interviews and the normal distractions of your day, you will reinforce for your candidates and yourself the seriousness about the opportunity.

The courtship

Solicit input from a range of stakeholders (within and external to your business unit). Carefully choose who will represent the company to the candidate. We tend to have positional perches, unconscious biases, filters and agendas, which can cloud our perception of the candidate. Where your stakeholders validate your perspective then the engagement process can proceed with confidence.

Further interactions, to facilitate greater engagement with the candidate and assist in selling the firm and aligning expectations may include:

- A kick-off meeting by a senior executive
- At least one main social event with potential colleagues
- Interviews and presentations (in groups and/or one-on-one)
- Relaxed, informal meetings with recent hires to get their impressions of your company
- Candidate meetings with HR to ask questions about the on-boarding process or career development
- A memorable last impression and quick follow-up.

After the initial meeting, it is essential there is a plan to asking questions, not just free-flow dialogue.

Questions should be structured to delve into what the candidates have done in the past and what they have learnt from these efforts. Patterns of past behaviours, combined with the appropriate motivation to achieve, are good predictors of future performance.

Carefully observe employees in your organization who excel in the role you're looking to fill. Identify their key behaviours, determine the relevant competencies candidates should exhibit and ask questions surrounding this.

Get more creative with your questions. Ask questions in which the answers could challenge your first impression.

Be quick to notice and slow to judge

After the interviews are completed, ensure there is a post interview discussion to consolidate findings and determine the best way forward. This should be done diligently, as with our project opportunities, go versus no-go decisions on candidates should be made swiftly.

Keeping in touch

The hiring process can be likened to a corporate version of a courtship in which emotionally charged parties put their best feet forward in nurturing potential employment relationships. We know of too many recruitment efforts that failed for lack of attentiveness to the feedback process. Candidates are customers too, and they deserve prompt feedback. The sooner you close the loop, the sooner you can move on to even more substantive discussions with your favoured candidate. Nominate one person to stay in contact with the candidate through the recruiting process.

Formulating the invitation

> It doesn't interest me to know where you live or how much money you have. I want to know if you can get up after the night of grief and despair weary and bruised to the bone and do what needs to be done.
> It doesn't interest me where or what or with whom you have studied. I want to know what sustains you from the inside when all else falls away.

The above excerpts from *The Invitation* by Oriah (2017) express that an invitation needs to go below the surface, and really find out what an individual's values, beliefs and actions are. In a recruitment context, formulating an invitation to join the firm should take into consideration these underlying values and interests. Therefore, the engagement process should determine intrinsic motivators – that is, why the individual would engage with you for their own personal satisfaction.

Following the above process should give you an in-depth understanding of the candidate's drivers to making a new move and therefore putting together a winning proposal should be relatively straightforward. In addition to the base salary, and variable pay incentives, gather an understanding of the intangibles and perks that may lure your candidate. Pay due consideration to job titles, opportunities for continued learning, international assignments, mentors, or salary-sacrifice options.

Personalise the delivery of the offer. Invite the candidate to a final meeting and walk the candidate through the proposal. When the offer is accepted, ensure you stay in touch with them through their notice period. For some people this is quite an emotional and difficult time, so ensure you are there to support them, and reinforce they have made the right decision.

On-boarding

On-boarding is far more intimate than induction. On-boarding is analogous to the first day at school. We all remember that day. The on-boarding process and orientation to the new company should begin once the individual has accepted the new position. On-boarding the new recruit is a shared responsibility and a strategic approach increases the likelihood for success. Proper management of the on-boarding process can greatly enhance the chances of a smooth and productive relationship between the new recruit and the organization.

Integration

Key elements to the Integration process include:

- Examine and talk to previous employees who have had earlier success during the on-boarding process. Be open to how we could have made a smoother transition for them.
- Ensure attendance to company inductions and information sessions.
- Use some fanfare. The start of any new recruit represents an opportunity to advance the cause, celebrate the fishing expedition and gain support. By getting employees ready it demystifies any confusion and uncertainty. The fewer details you leave to chance, the smoother the transition will be, and the more likely the new recruit is given the opportunity to succeed.
- Press announcements and articles in the company newsletter for a strategic recruit can highlight how their background fits with corporate strate gic direction. Properly done, these can generate enthusiasm and excitement about the new recruit.
- Use a buddy or mentor. Often, members who participated in the recruitment process fill the role most comfortably. Also, those with some tenure in the company or status can often provide counsel and support as the new recruit navigates through new challenges and relationships. The goal here is an honest broker, a candid sounding board, and a keeper of confidences with whom the new recruit can be fully candid.

Existing staff expectations can affect the success and performance of a new recruit. Involving staff or communicating to them during the recruitment process can facilitate buy-in and acceptance.

Transforming our recruitment mind-set

If you see an excellent candidate and assessment and reference checks validate this, make the decision, don't procrastinate and continue to ask for comparisons. In today's market we cannot afford to wait. Candidates are in short supply and the fishermen who hesitate will lose the fish and this may well mean starting the fishing expedition over again.

Suggested actions for encouraging staff to develop the above mindset in recruitment are listed in Table 14.4.

Table 14.4 Steps for transforming our recruitment mindset

ID	*Activity*	*Action*
1.	Undertake research on actual recruitment experiences from: Recent hires Candidates who have declined an offer Recruitment agencies.	HR Reps
2.	Ensure all frontline personnel attend recruitment and selection workshops.	HR reps
3.	Capture positive examples of the recruitment experience and promote them to all staff.	HR reps
4.	Enhance marketing material used throughout the recruitment process.	Marketing and Regional HR
5.	Ensure buddies and mentors are allocated during the on-boarding process and ensure they are aware of their role.	Line managers and HR
6.	Ensure staff are aware of current employment opportunities to further promote the firm's induction and on-boarding.	Line Managers and HR

Measurement

The following indicators may be used to assess that a transformation in the recruitment mindset has occurred:

- An increase in the number of candidates put forward from internal referrals
- Increased involvement of a variety of staff in the recruitment process
- An increase in the pool of quality candidates available for selection
- The amount of candidates that accept an offer made to them is high
- On-boarding survey results are positive showing successful integration
- Greater awareness from staff of new starters
- Increased promotion of strategic recruits across the region.

15 Business systems and processes

This chapter describes the attributes of the organisational business systems and processes. Chapter 8 identified that a PMO is used as the 'white van' knowledge centre of project management practices and processes. This chapter outlines the requirements for a lean and efficient Business Management System (BMS) that needs to be continually refined and updated to address the real operational risks.

15.1 Business management system

Per our strategy map in Chapter 2, a sizeable firm needs the business systems and processes to support the delivery of projects through effective planning and managing, adding value to the provision of these services while mitigating risks through flawless execution.

With the evolution of many professional services firms over many years, their Business Management System (BMS) has become an amalgam of forms, procedures, and processes which have often become unwieldy, a significant burden, and a handbrake on the business.

A lean and efficient BMS includes a hierarchy of controls, summarised as follows:

- Level 1 mandatory requirements (Must): requirements that must be complied with per the delegation of authorities and subject to audit that if breached would be subject to three warnings under a just culture
- Level 2 procedures (Shall): requirements that are preferred procedures but may be adapted to suit the project-specific requirements
- Level 3 guidelines (May): requirements that are provided as guidance or practice notes to assist project managers, commercial and the wider project team.

All staff need to be educated on the Level 1 requirements to understand that these requirements are not negotiable, without approval or a waiver. The firm's most valuable asset is its reputation. Safety and ethics are the non-negotiables and the BMS needs to clearly state the mandatory requirements.

As explained in Chapter 7, a Project Management Office (PMO) is connected to, but separate from the BMS and this highlights the distinction between project and corporate governance.

15.2 Building information management

Building Information Management (BIM) is more rightfully known as an Owner's information management system. It is intended to capture all project and asset information, documentation and data digitally across the project life cycle from planning, design, construction to operations and maintenance. It is often also referred to as digital engineering, such as 3D models, 2D drawings, specifications, visualisations and other project data.

The beneficiaries of BIM are:

- Stakeholders: can visualise the facilities in 3D and walk through the assets during the planning and design development phases, rather than 2D drawings supporting operability
- Constructors: can visualise the construction of the works for all workers to understand how the works are to be constructed safely. Also for capturing testing and commissioning data and supporting achieving technical completion and systems assurance
- Suppliers and vendors: can use the 3D models for shop drawings and fabrication
- Owners: BIM technologies, process and collaborative behaviours will unlock new, more efficient ways of working at all stages of the project life cycle. Also Owners can manage their assets more efficiently with better information and data.

On the basis that Owners need to spend money to save money, BIM places an additional burden on professional services firms to design and capture information in the 3D models. The concept of 4D BIM has become a buzzword in recent years. This equates to the use of BIM data to analyse time, beyond this is 5D, which includes cost management, and 6D for facilities management (FM) purposes.

Many government agencies are mandating the use of BIM. By example, the UK government (HM Government, 2016) has mandated level 2 BIM on all public infrastructure projects from 2016. Level 2 BIM requires all project and asset information, documentation and data to be electronic, which supports efficient delivery at the design and construction phases of the project. At the design stage, designers, Owners and end users can work together to develop the most suited design and test it on the computer before it is built. During construction BIM enables the supply chain to efficiently share precise information about components, which reduces the risk of errors and waste.

Level 3 BIM will enable the interconnected digital design of different elements in a built environment and will extend BIM into the operation of assets over their lifetime. It will support the accelerated delivery of smart cities, services and grids.

Per our strategy map, professional services firms need to be at the forefront of technological advances. However, this can be a significant investment in the learning curve with the application of these new technologies.

16 How to increase the chance of project success

> There is nothing noble in being superior to your fellow men. True nobility lies in being superior to your former self.
>
> (Ernest Hemingway)

This chapter consolidates all the organisational-management and project-leadership learnings in the previous chapters into a complete summary of how to increase the likelihood of a successful project, or portfolio of projects.

16.1 Overview

The maturity of project management and governance competencies varies by jurisdiction and market sector. Project Owners need to review the risk allocation in the procurement supply chain to ensure project success for all project participants. Project Owners should consider the contracting methodology that best matches the risks of the project.

Risk management embodies an organisational and project culture of prudent risk taking. Unfortunately there are too many examples of major projects that have failed to achieve the project's objectives, with delays, overspend and operational issues. Although the construction industry (and insurers) may have borne the brunt of the project losses, these costs must come back to project Owners on future projects.

Risks can be manifested on major projects by happenstance or chance, or they can be mitigated or managed through a structured process of risk management. By understanding the root cause of the risks to project success, all of the project stakeholders will gain greater certainty of achieving the project objectives.

Project Owners should invest in developing close working relationships with contracted parties to earn mutual trust and address risks collectively. Rather than simply passing on all or most of the risks to the construction parties, it would be preferable to establish a more equitable risk allocation, based on the party best able to control the risks. This implies closer working relationships between Project Owners and contracted parties working as an integrated team with common project objectives and incentives.

This chapter utilises risk-management principles and practices to determine what makes a project successful, utilising lessons learned from a portfolio of major projects, be it Design & Construct, Engineer, Procure & Construct, or Collaborative contracting form of project delivery. These experiences suggest that the challenges to major projects and high-exposure risks are often non-technical in nature and not often addressed in conventional risk management practices. The human factors in developing a risk-based culture in the project management and governance functions are often the key ingredients to the assurance of project success.

Effective risk management provides a competitive advantage to the contracted parties and represents enhanced value-for-money to the Project Owner by not exhausting risk contingencies or compromising operational requirements.

16.2 A metaphor

In Greek mythology, Icarus is the son of the master craftsman Daedalus, the creator of the Labyrinth. Often depicted in art, Icarus and his father attempt to escape from Crete by means of wings that his father constructed from feathers and wax. Icarus's father warns him first of complacency and then of hubris, asking that he fly neither too low nor too high, because the sea's dampness would clog his wings or the sun's heat would melt them. Icarus ignored his father's instructions not to fly too close to the sun, whereupon the wax in his wings melted and he fell into the sea. This tragic theme of failure is at the hands of hubris.

One of the key risks to project success is hubris, excessive self-belief, ego and unconscious biases. In an era where most projects fail, the Heathrow Terminal 5 was on-schedule and budget. However, it failed to achieve its operational performance requirements on the opening day. A story of hubris by Tom Brady (2010) on the Heathrow Terminal 5 project reinforces this view. High-exposure risks are often non-technical in nature and this chapter explores these risks to achieving project success.

16.3 Setting the context

Efficient delivery of infrastructure is the hallmark of a well-functioning economy. The Australian Government, Productivity Commission (2014) inquiry into funding public infrastructure highlighted that there is evidence of significant increase in the costs of constructing major public-infrastructure projects in Australia. While poor procurement practices, escalating labour costs, industrial relations, and flat-lined productivity are often considered the cause of over-runs, the commission concluded there is no single input that has played a decisive role in cost increases.

The KPMG 2015 *Global Construction Survey* (KPMG, 2015) reported that 53 per cent of Owners suffered underperforming projects in terms of budget and schedule performance in the previous year. For energy and natural resources the failure rate was 71 per cent and for public infrastructure 90 per cent. With 14 per cent of the respondents from Australia and New Zealand, a number of projects compared favourably in terms of project-delivery processes; however, there were opportunities to improve risk allocation and project management in the maturity curve.

In investigating productivity of the construction industry, McKinsey (McKinsey & Company, 2015a) reported that 98 per cent of megaprojects failed to achieve budget, schedule or specification requirements and there are a variety of reasons for the poor performance. It recommended that better project management and technological innovation can improve the chances of success. Causes for poor project performance were highlighted as poor organisation, inadequate communication, flawed performance management, contractual misunderstandings, missed connections, poor short-term planning, insufficient risk management, and limited talent management.

At their annual conference, Independent Project Analysis (Independent Project Analysis Inc, 2015) presented the results of a study of over 3,700 projects, reporting that 67 per cent of megaprojects failed to achieve their objectives. The research suggests that the project sponsor (aka Owner) is responsible for shaping the front-end definition on complex projects and setting the project on a pathway of success or failure from the outset. They opine that most major projects derail well before construction commences. The onus is on the Project Owner to perform sufficient front-end analysis with suitably competent professionals to shape the project for success.

Infrastructure Australia commissioned the Caravel Group (2013) to undertake a survey of industry and government agencies to determine project success rates. Their survey highlighted an inconvenient truth that 48 per cent of projects failed to meet their baseline time, cost and quality objectives and there had been little performance improvement in recent times. They concluded that the major cause of project failure was the lack of project governance.

The method of procurement of major projects by the Project Owner is key to determining project success. There is a project myth that pure alliance or collaborative contracts develop soft target estimates and as a consequence competitive alliances or competitive early contractor involvement has been introduced. This myth is dispelled through the research by Derek Walker of RMIT (Walker, 2015) showing that of 60 pure alliance contracts undertaken in Australia, around one-third exceeded the target estimate or the planned schedule. The better success rates of alliance contracting can be attributable to better project shaping through the project-development phase.

In 2012, the Victorian Auditor General (Victorian Auditor General's Office, 2012) undertook an audit on managing major projects, which found that the responsible government agency was not managing major social infrastructure projects effectively, highlighting a lack of basic governance, major deficiencies with the management of internal contracts, inadequate project management, a lack of relevant and appropriate performance measures, and not relevant data to assess how it is performing.

In response to many of these industry challenges, Engineers Australia convened a conference titled Mastering Complex Projects (Engineers Australia,

2014) seeking to address the causation for poor success rates. Complex projects with budgets greater than $500 million have a much higher failure rate than their more straightforward counterparts. In some sectors, up to 75 per cent of complex projects fail to achieve their budget, schedule or operational performance requirements. This represents a vastly inadequate return of investment for both private enterprise and taxpayers. Their white paper highlighted that projects fail for many reasons, such as lack of communication of stakeholders, critical skills and knowledge gaps for key personnel, poor conceptual planning, insufficient implementation of project controls and risk management, and ineffective transfer of lessons learned from similar projects.

There is also significant other international evidence that mega and major projects are failing to achieve their project objectives, providing significant burden on the industry, most notably by Bent Flyvbjerg (2014). While many Owners may consider that the contracted parties have suffered the losses and this doesn't impact taxpayers, the impact of these costs will be borne by future projects in either increased risk contingencies or increased margins.

There is a clear value proposition that we need to improve our project success rates to more efficiently deliver the public infrastructure that is vital for our social, economic and environmental well-being.

16.4 Increasing the likelihood of success

This section provides a unique perspective on engineering management risks from involvement in many major projects, joint ventures and collaborative contracts with Owners and contractors. The terms used in the chapter have been purposely selected and are explained as follows.

In this context, engineering management is the fulcrum between design and construction, addressing the Owner's requirements. The author suggests that the management of engineering is a lost skill and the industry should consider re-skilling professionals to be adept at managing the needs and wants of the Owner, designer, constructor and related stakeholders.

In the title to this chapter, 'how' implies that there is a structured methodology or approach to achieving project success. Chance is used to highlight that risks are not manifested by happenstance or luck but should be considered through the application of knowledge, expertise and experience during the planning, procurement and delivery phases. Similar to the approach to safety, that accidents are preventable, so too are the other risks. In this context, success means achieving the project objectives for both cost and non-cost performance metrics.

The above contextual setting suggests there are a multitude of causal factors to projects failing to achieve their objectives. An article by Graham Scott of Thinc Projects (Scott, 2013) suggests that the impact of human relationships is a key contributor to project success. He opines that projects generally fail because of human behaviours, not technical issues.

Project management is an oxymoron. We don't actually manage projects, we manage people to achieve the project objectives. This section focusses on the people and culture perspectives with reference to the systems and processes. With the scale and complexity of many public infrastructure and resources projects, it is to be expected that risks and challenges will arise during the journey. The best project teams are not fair weather sailors, but rather teams that can rise to the challenges that the elements present. Projects are essentially a series of problems to be solved or risks to be managed.

The engineering and construction industry has been through a transformation in recent years in the pursuit of zero-harm safety performance, seeking to catch up with other industries. This aspiration should be applied to all risks including ethics, planning, delivery and reputational risks. In the inquiry into the RAF Nimrod aircraft crash, the Hon Charles Haddon-Cave (2005) stated that 'questioning is the antidote to assumptions that so often incubate mistakes'.

The proposition is that human factors are a major cause for major projects not achieving their objectives. The following sections discuss these causal factors, focussing on project leadership, culture, collaboration and relationships, tendering legacies, risk management and assurance of project success.

16.5 Project objectives

Project success is defined as achieving or exceeding the project objectives, often expressed as the iron-triangle of budget, schedule and scope requirements. Steven Covey (1989) stated in his study of the seven habits of highly effective people to begin with the end in mind. This is relevant to the setting of project objectives, or the end-state in military terms, to clearly articulate and communicate the purpose and what would success look like (or feel like) at the end of the project.

Functional project teams have small numbers of people with complementary skills who are committed to a common purpose, goals and approach for which they hold themselves accountable. In simple terms: 1 + 1 = 2. The project objectives are defined as what success would look like at the end of the project, defined in both cost and non-cost terms. Non-cost objectives may be safety, ethics and reputation, quality, excellence, innovations, personal development, industry legacies, sustainability, the stakeholders' and the Owner's satisfaction.

However, there are many examples where high-performance project teams can be formed that meet the conditions of functional teams but have members who are deeply committed to one another's personal growth and professional success. High-performance project teams have a group identity, with high enthusiasm and energy levels, personal commitment and are results oriented. They exhibit a best-for-project approach, project culture and team spirit. They normally have strong project leadership from a management and governance perspective. They normally achieve break-through performance, as outlined in the project-alliancing practitioners guide (Victorian Department of Treasury & Finance, 2006). That is: 1 + 1 > 2.

There are many sporting examples where high-performing teams have achieved the improbable. High-performing teams respond to stretch targets. The targets may be considered improbable and at first thought not obviously achievable, but they are not impossible. Alchimie (Alchimie and DLA Piper, 2003) contends that achieving the improbable stretch targets implies identifying, mitigating or managing the project risks and harnessing innovations and opportunities.

Financial performance is an outcome of doing the right things, performing the work as planned and managing risks, both threats and opportunities. The objectives also need to define the behaviours that will produce the target outcomes. Like a sporting team, it is the planning and tactics that create the results on the score board. Setting the right objectives, with stretch targets and getting everyone in the project team to buy into them, will increase the chance of project success.

16.6 Project leadership

Leadership and culture are intertwined. Great project leaders create a great project culture. And a great project culture fuels the project leaders with the energy to exceed the project objectives.

However, culture is not solely a leadership responsibility, it is everyone's responsibility to uphold the culture. Abraham Lincoln stated 'it takes many good deeds to build a good reputation, and only one bad one to lose it'. Project culture can easily be eroded by toxic attitudes, negative and critical perspectives, individualism and self-interest, often referred to as 'terrorists in the camp'. These people often lack confidence, exhibit self-interest, negativity, avoidance, internal competition, and blame. The leadership team need to address these inappropriate behaviours, no matter how strong their technical competency. We often tolerate poor behaviours because of individuals' technical competency; however, avoiding these difficult conversations will not increase the chance of project success. Project leaders must set the standards and ensure that people do not remain in the team if they exhibit inappropriate behaviours.

Using activator-behaviour-consequence (ABC) analysis to adapt behaviours, leading teams requires an understanding of the activators or triggers that will create the appropriate behaviours and the consequence of those actions, under a just culture. Positive activators and reinforcement of behaviours is more likely to achieve better outcomes than draconian criticism. The interaction of key behavioural indicators (KBI) with key performance indicators (KPI) is illustrated in Figure 16.1.

While individuals are selected predominantly for their technical skills and expertise, their personality may not permit them to work in a team environment and they should not be part of the team. It is important that the standards of the team are upheld. To address the risk of poor team formation, psychometric evaluations such as the Human Synergistics (Human Synergistics, 2017) circumplex model can assist. The constructive styles are highly effective and promote individual and collective performance. In contrast, the non-constructive aggressive, defensive, passive or aggressive styles have an inconsistent and potentially negative impact on performance and consistently detract from overall team effectiveness.

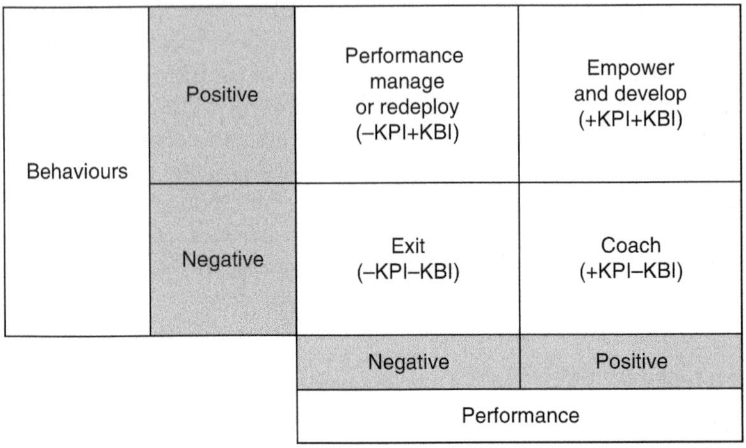

Figure 16.1 Performance behaviours matrix

The rasion d'être of the project leaders is to create a risk-aware culture. That is, ask the right questions, consult with the right people and challenge the norm. In *Strategy of the Dolphin* (Lynch, 1990) the authors developed a useful metaphor which is appropriate for the construction industry. The dolphin leadership style (compared to sharks and carp) is driven by a deep sense of vision and pragmatic realism of risks (both threats and opportunities) that enable them to mobilise the project team towards a common goal.

While project reporting needs to be based on quantitative factual data, the qualitative narrative (interpretive reporting) by the project leader provides the underlying vibe or culture of the project. Their narrative is insightful and purposeful for the project governors to gain a sense of confidence that the project team is working cooperatively and collaboratively to address risks. In essence, the enlightened project leader is a risk-taker, prepared to take risks as part of a deliberate process, where those risks are susceptible to mitigation or management in such a way as to achieve net positive project outcomes.

16.7 Risk culture

Human Synergistics (Human Synergistics, 2010) suggests that culture can be defined as shared values, norms and expectations that govern the way people approach their work and interact with each other.

When mobilising a project team, Jim Collins (2001) contended a project leader needs to get the right people on the bus, the wrong people off the bus, and the right people in the right seats. On-boarding is an important factor and is often overlooked, given the need to mobilise expeditiously to meet the deadlines. On-boarding is much more than an induction. Too much haste and not enough speed often means that the team is not formed as well as could be expected to maximise the synergies and potential of the team. Lack of investment in the on-boarding can often lead to 1 + 1 < 2.

Whether a sporting team or a project team, there need to be standards (aka a code of behaviours). The AFL team the Sydney Swans, for example, are known for their adherence to standards. The leadership team set the standards, with zero tolerance for non-compliance.

Developing or transforming an organisational culture takes months. With new project teams, this needs to be achieved within a matter of weeks. What creates the culture we aspire to? What drives the behaviours that create the project culture? What are the ingredients or causal factors that drive the desired outputs?

There are many causal factors to creating a high-performing project culture, including the personality of the project leader and the management team, role clarity, optimism and confidence in achieving targets, regular team talks and open communications, and constructive behaviours. Communication influences the behaviours that create the desired culture.

Alliance contracting played a significant role in the development of collaborative or relationship-based contracting. The industry has developed beyond singing *Kumbayah* or undertaking team-bonding activities, to understanding the need to take time to get to know each other and define how best to relate. In other words, undertaking a stocktake or due diligence of each individual's strengths and weaknesses to be able to glue together the knowledge, skills, experience and attributes (KSEA) of the collective group. The project leader who can harness the combined skills, expertise, and energy will develop a project team that will 'smash' the targets. The project leader is the custodian of the project DNA.

For large-scale and complex projects, there will be many challenges to be addressed. There will be internal risks and external risks. Internal risks such as an under-cooked tender (which we will examine later in this chapter), lack of collaboration, lack of team harmony, misaligned corporate and individual interests. As well as external risks such as a difficult Owner, pedantic and preferential contract administrator, stakeholder relations, protracted approvals, utility services, contamination, unforeseen latent conditions and inclement weather.

On many major projects, alliances or joint ventures are formed to share the risks. In an overview of strategic (organisational) alliances (Forbes, 2002), the author investigated the risks and problems facing strategic alliances, and found that cultural clash is one of the biggest problems that corporations face. Cultural disharmony can arise from language, diversity, ego, chauvinism, different attitudes, and mindsets. Considering a major project as a virtual organisation, this is also true for projects, with an eclectic amalgam of knowledge, skills, attributes, experiences and personalities.

Ed Schein (1985) stated that if you set out to change the culture, you end up in fog. If you set out to change behaviours, then you'll change the culture. In today's environment, our project leaders and governors need to transform the culture from being risk-averse to being proactive risk managers. The two extremes are being risk-averse by not taking risks and being blindsided, compared with taking risks but not managing them, which is tantamount to gambling.

Having a strong risk culture does not mean taking less risk. In fact, managing the people-side of risk (McKinsey & Company, 2015b) with an effective risk culture may mean taking significantly greater risks, but with appropriate controls. By example, the brakes on our cars actually make us drive faster, not slower. Proper controls can provide a change in mindset that will create a culture of being risk-aware, responding to challenges and driving innovation that will increase the chance of project success.

16.8 Collaboration

Relationship management is a risk and should be managed accordingly. The importance of relationships is often misunderstood by technical specialists and is often out-sourced to a facilitator, rather than considered as a core leadership function in the belief that it occurs naturally and does not follow a process that can be monitored and controlled. Communication, collaboration and effective relationships are key ingredients to project success.

The chance of project success is increased if the relationships with related stakeholders are sound and effective based on trust and respect. One of the key risks to project success is expeditious approval of documentation by the Owner and related stakeholders. The stakeholders must also be on the bus for the journey. The project objectives will not be achieved by the project operating within a bubble or vacuum. The project team will have many touch points with the related stakeholders, with every interaction an opportunity to build that trust and confidence. Again, for the technically minded, the David Maister (2001) trust equation can be adapted as:

$$\text{Trust} = C \times R \times Q$$

Trust is developed through credibility (C), reliability (R), and quality (Q) of the relationship with a common interest in project success. Credibility is the track record and experience of the project team. Reliability is earned by delivering quality documentation on time, being empathic to the reviewer's resources and operational constraints.

One of the biggest challenges in fast-tracked delivery is expeditious review and approval by the Project owner and third parties. Documentation is more likely to be approved expeditiously if it is complete and correct without errors, omissions and is not submitted in pieces, where the complete picture is not evident (Engineers Australia, Queensland Division, 2005). These factors are enhanced when engaging relationships with all third-party stakeholders.

Fast-tracked approvals need to be structured in a manner so that the approving authorities appreciate they are a valued part of the project process and not the handbrake to achieving the programme. Again, the chance of project success can be increased when the approving authorities feel they are an integral part of the workflows and contributing to the project success.

16.9 Project governance

Governance is defined as the processes implemented to provide management oversight, guidance and compliance to policy and procedures. There are two types of governance; corporate and project governance. A project is at-risk when individuals do not understand when to wear which hat and misunderstand the differentiation. Corporate governance relates to authority, direction and control. Project governance requires knowledge, skills, experience and attributes (KSEA) that relate to project risks.

The Caravel Group (2013) report referred to earlier suggestions that delivery of project governance is generally highly dysfunctional. One of the causal factors for this could be self-interest and hubris and individuals misunderstanding the difference between project and corporate governance. In simple terms, individuals view their role as representing their organisation's interests and not the collective interests of the project. Project governors need to govern upwards to their respective organisations as well as downwards to the project team and sideways with related stakeholders.

There needs to be a governance plan or charter to guide the conduct of the participants on a project board, steering committee or project control group. And the plan needs to be measured at least on a quarterly basis. For larger projects, with large and eclectic participants, the appointment of an independent project governor may address this risk. The role of the independent chair would be to conduct the project governance meetings in a planned and managed manner, without bias or favour.

Project governance committees need to be run with the same rigour as statutory boards (ASX Corporate Governance Council, 2010). One of the core principles of the board is to recognise and manage risk. To increase the chance of project success, the project governors need to have the requisite knowledge, skills, experience and attributes (KSEA) and engaging personality to perform the role and be held accountable for their actions. Their appointment should be based on merit, not on position. The tone of the project is set from the top down.

16.10 Structure and people

The Project Management Body of Knowledge (Project Management Institute, 2013) defines a project as a temporary endeavour that has a defined beginning and end in time, and therefore defined scope and resources. As a temporary endeavour, resources are normally sourced from a variety of sources that need to gel together in a short period of time. A project organogram is an important document for harnessing the skills and expertise of the project team. The risk arises when personnel are not on-boarded properly and the team will not be functional or productive. The organogram should not be considered as a 'wiring diagram' of command and control, but rather a network of relationships. The characteristics of great working relationships are trust and respect. The on-boarding process is therefore of vital importance and often not well considered, given the haste of mobilisation. Role clarity is of vital importance not only for understanding of individual accountabilities but also those of the people to interact with and to be relied upon.

Projects do form their own unique culture, often sourced with personnel from disparate participant organisations. The term often used is 'leave your corporate hat at home'. As explained in Chapter 3, many project teams suffer from 'Stockholm syndrome', the term applied after a group of hostages sided with their terrorist captors. Whilst a unique project culture should be formed, it should not be elitist or show excessive hubris having disregard for the alignment with the respective organisational culture and business processes. Again this highlights the need to align project governance with corporate governance.

We are all familiar with the five phases of project development from forming, storming, norming, performing, and adjoining. It should be expected that all newly developed project teams will have a storming phase. Dysfunctional teams do not respect one another, are not willing to accept differences in a constructive way, do not have a commitment to a common goal, are not willing to be accountable, lack clarity of role, have low standards, have an avoidance behaviour, exhibit internal competition with a blame culture driven by individual egos. This is often referred to as the 'seagull syndrome', fighting over the chips for individual gratitude.

To address this risk, it is important that the development of the organogram is properly communicated with supporting position descriptions, management plans and an interface matrix for clarity of the interactions. A responsibility assignment matrix or RACI matrix (Project Smart, 2017) describes the participation by various roles in completing tasks or deliverables for a project, the interaction with other disciplines or functions, and the reliance on other project team members.

While draft management plans may be sourced from past projects or a project management office, they need to be specific to the project and owned by the responsible person. Often management plans need to be approved by the Owner as a condition precedent. Management plans need to be considered as the 'music that keeps the orchestra in harmony'. Too often management plans are verbose with motherhood statements and not relevant to the specific project risks. They should be short and concise documents that are live and relevant for the on-boarding and management of the project team.

16.11 Project management

While the previous chapter has focussed on the cultural and leadership factors, effective project management is the foundation for achieving the project objectives, addressing the iron-triangle of time, cost and quality. Too often we only achieve two of these requirements.

The project management plan (aka project execution plan) is the 'bible' for the project. Similar to position descriptions described above, it should not be overly verbose or filled with motherhood statements from the business management system. It needs to be project-specific and actively communicated to all personnel. The plan needs to address the risks not able to be mitigated in the contract. The project management plan should be approved by the governance team. Similarly the project management team need their own subordinate management plans for their area of responsibility that are specific, relevant and communicated effectively.

On the commencement of the project, the objectives define what success will look like. Along the journey, it is important to measure progress and continually forecast the route to completion, in a no-surprises approach. Peter Drucker (1954) contends that what gets measured gets managed. Effective project controls are paramount for increasing the chance of project success. There is much literature about earned-value measurement; however, accurately assessing progress and accrued costs often requires professional judgement. Many personnel, proud of their work, can be overly optimistic in their assessments. This is often another unconscious bias.

In terms of measuring progress and alerting risks, we use the term values-based assurance. That is, there is an assurance that their part of the project is on-track, unless advised otherwise, in a no-surprises approach. A core value of most organisations and projects is integrity. Not reporting an emerging risk or over-reporting progress should be regarded as a breach of integrity, with due consequence. Integrity is like the internal sat-nav that guides all project personnel to do what is right.

Similar to ethics, the culture of projects needs to be such that personnel feel comfortable accurately reporting progress and risk. The project leader should create an environment of sanctuary for early notice of risks and, conversely, consequences, should risks be hidden or buried, in a just manner.

These assessments then lead to an informed decision about drawing down on the project contingency. Contingency is not a war-chest, or an ATM. It is a project cost that is expected to be expended. Contingency management is a core part of project management and advising the governance committee on the likely forecast at completion. It would be expected that contingency would not be drawn upon each month on a linear basis. The conversation with the governance committee needs to focus on what is the appropriate level of contingency for the forecast risks.

Contingency drawdown is not likely to be linear from start to finish. It is likely to be drawn upon in the first quarter and then again in the last quarter. Astute re-forecasting will validate the required level of contingency.

16.12 Innovation

The definition of innovation is to introduce something new, such as a new idea, a new approach or a new process to achieve a better outcome than business as usual. This would be measured with a better key result area performance score. Innovation is often regarded as the engine which can introduce construction economies and advance labour productivity. Project leaders should be committed to creating a culture of innovation and recognising and rewarding personnel that devise a new way of doing something that yields a tangible benefit to the project.

Innovation lies at the heart of collaborative contracting and when implemented will yield improved value-for-money for the owner. The engine-room provides the forum for the project team to collaboratively devise better ways of performing the works. An innovation can be anything from a big-bang great idea, to a minor change to a traditional process, such as identification of an inherent risk, or an insight that is not evident to the team. Like any change, a small innovation at a critical stage can have a significant impact on the likelihood of success.

Examples of innovations are ingenuity, collaboration, problem solving, productivity, communication, risk awareness, flexibility, mindfulness, interface coordination, timeliness, and alerts. Innovation does not mean re-inventing the wheel or debating issues ad nauseam. It does mean the team is empowered to challenge existing paradigms. Most project teams have extensive intellectual horsepower and experience by embracing a culture of innovation, they will yield significant benefits in both cost and non-cost performance metrics.

However, innovations must be implemented in a timely manner. Late innovation can lead to change and this can infect a project like a virus. Creating a culture of timely innovations will increase the chance of project success.

16.13 Continuous improvement

As an industry, we need to learn the hard-earned lessons from past projects to capture the project knowledge and learn from what went well, what didn't go well, and what should be done differently. As projects are temporary endeavours that are linear, with a start and end point (unlike a portfolio or programme or works), there is often little handover of knowledge from one project to another. Many firms seek to capture lessons learned in their project management office for implementation in the start-up of the next project. However, there are barriers to incorporating experiences from the past. And the key barrier is the hubris of the new project team that take little regard for past learnings, thinking they know better.

While the legal fraternity preclude the industry from sharing mistakes, again it is the human factors that prohibit the sharing of project knowledge in the transition from one project to another with a new project team. It is proposed that there be a lessons-learned exercise on lessons learned. Studies show that only 10 per cent of people who have had heart bypass surgery or an angioplasty make major modifications to their diets and lifestyles afterwards.

Often a lessons-learned exercise is undertaken at the end of a project. At this stage, many personnel have been demobilised and the project memory has faded. It is recommended that the lessons-learned knowledge capture start from the commencement of the project and one of the project management team be responsible for capturing the pros and cons and the opportunities for improvement. Capturing lessons learned is an integral part of risk and opportunity management. Thinking about innovations and opportunities for improvement and capturing these progressively will increase the chance of project success for both current and future projects.

16.14 Tendering risks

Many of the fundamental project activities can be directly related back to the awareness, identification and management of risk. To illustrate how this discussion can be applied, this section considers one of the key risks faced by delivery teams; design growth arising from a preliminary tender design. This means growth of the tendered quantities from a preliminary tender design to the actual constructed design.

To set the context, on many projects, the Owner's reference design is well-developed and the designer is engaged to develop this into a design suitable for tendering by an experienced and competent constructor. The tender design is a preliminary design based on limited inputs such as survey, geotechnical investigations, equipment vendor information and construction methods. It is not a complete design, and as such the tenderer needs to consider known risks, known-unknown and unknown-unknown risks. A risk contingency is required in the tender price.

An overly conservative tender design will be non-competitive. The project team needs to determine the reliability of the tender design to:

- accurately estimate the direct costs, and
- develop into a detailed design for fast-tracked procurement, approval and construction purposes.

Public infrastructure projects are highly competitive. Innovative or alternate designs often provide a competitive advantage. However, these innovations are often less developed than the development of the Owner's reference design, representing greater risk.

The design-management methodology is the process whereby the designer works collaboratively with the constructor using every endeavour to minimise the direct cost of the works.

Most major projects are fast-tracked. The indirect costs are generally dependent on the duration of the schedule and the level of supervision. The team need to assess the constructability of the tender design and develop the most efficient sequence of the works and how the design can influence and impact the critical path.

It is to be acknowledged that the reliability of the tender design is highly dependent on the level of resourcing and effort required to prepare design information suitable for pricing by experienced estimators. Some elements require small amounts of effort, relying on benchmark data, whereas other elements require quite detailed analysis. In the current economic climate, tenderers will seek to minimise their tender costs, designing to a reduced budget, which can pose a significantly greater risk of design growth.

Design growth beyond the risk allowance does not necessarily mean the preliminary tender design was defective. There is a real risk that the risk contingency may be inadequate for the scale and complexity of the project. Many tenderers are now using probabilistic analysis of the direct costs to determine the likelihood of the risk contingency being exceeded using risk analysis, contingency determination and range estimating (AACE International, 2008).

To address this risk, the author suggests that the designers and constructors should work collaboratively to establish the risks (threats and opportunities) arising from an incomplete preliminary design and develop plans for mitigation, management, or pricing of the risk. This collaboration should be based on addressing the human factors as outlined in this chapter to estimate the quantum of the risk contingency, which in today's market can be the price difference.

In fast-tracked construction, the most successful projects have 'designed the project as tendered and constructed as designed'. Systemic change is the enemy of fast-tracked projects.

16.15 Risk management and assurance

As a consequence and legacy of the economic conditions, risk allocation is biased in the Owner's favour. A report produced for Infrastructure Partnerships Australia (Ashurst Australia, 2015) shows a growing trend that shifting risk to the delivery team is not necessarily representing value-for-money for the project, or the long-term sustainability of the industry. This highlights the need of the project team to be better skilled in the practice of risk management, not simply compliance.

Risk management should not be considered as the handbrake to project success. In fact, appropriate controls should in fact facilitate greater risk taking. Our cars would not be safe if we didn't have brakes. Therefore, we should be prepared to take risks, with the confidence we have the right controls. This is where project risk management can embrace enterprise risk management with the application of the three lines of defence in effective risk management (Institute of Internal Auditors, 2013).

The first line is the risk and opportunity management of the project-management team. Most projects are now adopting a structured process of identifying threats and opportunities, often in a facilitated risk-workshop environment. The risks are then considered based on the probability of occurrence and the consequence to evaluate the exposure. Monte Carlo probabilistic assessments are sometimes employed. Risk-management techniques are then considered and the residual risks understood.

The second line is independent peer review by experienced practitioners. This independent assurance is akin to quality assurance for the technical and non-technical risks. Acknowledging the proposition in this chapter that many major projects are derailed by non-technical risks, the independent assurance practitioner or panel should review both the technical and non-technical risks. This would review the people and processes top down from the project governance to the project management team and related stakeholders. The output of the review can range from suggested opportunities for improvement to intervention.

The third line of defence is structured quality control and financial audits for compliance with the accredited management policies and systems. In simple terms, risk assurance is earned through management, review and audit, tailored for the scale and complexity of the project.

Project teams are generally good at considering the technical risks. However, as explained above, it is identification and management of the non-technical risks that will increase the chance of project success.

Experience suggests that it is the risks overlooked or not properly considered that are manifested. These can be defined as blind-side risks. Similar to the blind spot driving a car, or the blind side in American football (Lewis, 2006), the obvious risks are often well-managed, and project teams are blindsided by the other risks. Enlightened project teams will have the wisdom and insight to foresee and pre-empt the threats before they manifest.

The Johari window (Ingham, 1955) provides a useful explanation of why we overlook these risks. There are things we know we know, things we know we don't know, things we don't know we know and things we don't know we don't know. While this quote was made famous by Donald Rumsfeld as the US Secretary of State, it is the opaque windows that pose the biggest risks to project teams (not black-swan events which are extreme outlier risks) (Taleb, 2007). Personal attributes such as hubris, arrogance, ego, excessive self-belief, optimism bias and unconscious biases are the blind spots that prohibit project teams from creating the awareness for identifying and managing these risks.

Similar to the diversity challenge, our unconscious biases are the greatest inhibitor to managing risks on our projects (Australian Public Service Commission, 2015). Another analogy is the risk of traffic forecasting of toll roads. The review undertaken by the Australian Government (Australian Government, Department of Infrastructure and Transport, 2010) showed that technical forecasting errors are a contributor, but also highlighted the non-technical risks of optimism bias referring to a systematic tendency for people to underestimate costs and overestimate benefits and strategic misrepresentation from pressures within organisations.

There are many that say the greatest leadership characteristic of successful leaders is awareness. That is, awareness of strengths and weaknesses, opportunities and threats to achieving the project objectives. In other words the project leader is effectively the SWOT commander.

If everyone in the project team considers themselves a risk champion, there is a collective appetite to manage threats and harvest opportunities. This would provide the confidence and assurance to take on the risks, not in a cavalier approach, but with due management, controls and governance.

16.6 Conclusion

This chapter presents the proposition that the root causal factor to project success is human factors. That is not to say that technical risks will not derail a project. However, given the maturity of the engineering and construction industry, there is strong reliance on the skills and expertise of the personnel involved and reliance on the quality assurance and independent review and approval processes.

Non-technical risks are not manifested on major projects by happenstance or chance. They can be identified, mitigated or managed through a structured process of risk management, if the project team is alert to them. Awareness is the key to managing risks. Project success is not serendipitous.

The human factors in developing a risk-based culture in the project management and governance functions are the key ingredients to assuring project success. Again as Ernest Hemingway stated, 'there is nothing noble in being superior to your fellow men. True nobility lies in being superior to your former self.'

References

AACE International. (2008). *AACE International Recommended Practice No 41R-08*. Retrieved from http://web.aacei.org/resources/publications/recommended-practices

Abrahamson, M. (2017). *The Abrahamson Principles*. Retrieved from Irish Times: www.irishtimes.com/news/crime-and-law/still-building-on-a-long-career-in-law-1.511540

Alchimie and DLA Piper. (2003). *Project Alliances an Overview*. Retrieved from www.alchimie.com.au/

American Institute of Architects. (2017). *Integrated Project Delivery: A Guide*. Retrieved from https://info.aia.org/SiteObjects/files/IPD_Guide_2007.pdf

Ashurst Australia. (2015). *Scope for Improvement 2015: Project Pressure Points – Where Industry Stands*. Retrieved from Ashurst: www.ashurst.com/en/news-and-insights/legal-updates/scope-for-improvement-2015-construction-and-infrastructure/

ASX Corporate Governance Council. (2010). *Corporate Governance Principles*. Retrieved from www.asx.com.au/regulation/corporate-governance-council.htm

Australian Accounting Standards Board. (2010). *AASB111*. Retrieved from www.aasb.gov.au/admin/file/content102/c3/AASB111_07-04_ERDRjun10_07-09.pdf

Australian Government, Department of Infrastructure and Transport. (2010). *Review of Traffic Forecasting of Toll Roads*. Retrieved from https://infrastructure.gov.au/infrastructure/infrastructure_reforms/files/Attach_A-BITRE_Literature_Review.pdf

Australian Government, Productivity Commission. (2014). *Public Infrastructure: Provision, Funding, Financing and Costs*. Retrieved from Australian Government, Productivity Commission: www.pc.gov.au/inquiries/completed/infrastructure/report

Australian Public Service Commission. (2015). *Unconscious Bias*. Retrieved from www.apsc.gov.au/publications-and-media/current-publications/human-capital-matters/2016/unconscious-bias#firstArticle

Australian Standards. (2006). *AS 4817 Project Performance Using Earned Value*. Retrieved from www.saiglobal.com/pdftemp/previews/osh/as/as4000/4800/4817-2006.pdf

Beaton Research and Consulting. (2017). Retrieved from www.beatonglobal.com/

Brady, T. (2010). *From Hero to Hubris – Reconsidering the Project Management of Heathrow Terminal 5*. Retrieved from Science Direct: www.sciencedirect.com/science/article/pii/S0263786309001446

Business Insider. (2016). *Cognitive Biases that Affect Decisons*. Retrieved from www.businessinsider.com.au/cognitive-biases-that-affect-decisions-2015-8?r=US&IR=T

Caravel Group. (2013). *A Review of Project Governance Effectiveness in Australia*. Retrieved from Infrastructure Australia: http://infrastructureaustralia.gov.au/policy-publications/publications/Project-Governance-Effectiveness-report-by-Caravel-Group-March-2013.aspx

Collins, J. (2001). *Good to Great*. William Collins.

Covey, S. (1989). *The Seven Habits of Highly Effective People*. Free Press.

Deloitte. (2016). *Cultivating a Risk Intelligent Culture*. Retrieved from Deloitte: www2.deloitte.com/lu/en/pages/risk/articles/cultivating-risk-intelligent-culture.html

Department of Infrastructure and Regional Government. (2015). *National Alliance Contracting Guidelines*. Retrieved from https://infrastructure.gov.au/infrastructure/ngpd/files/National_Guide_to_Alliance_Contracting.pdf

Drucker, P. (1954). *The Practice of Management*. Harper & Brothers.

Engineers Australia. (2014). *Mastering Complex Projects: Principles for Success and Reliable Performance*. Retrieved from Academia: www.academia.edu/33475220/Mastering_Complex_Projects_Mastering_complex_projects_Principles_for_success_and_reliable_performance

Engineers Australia, Queensland Division. (2005). *Getting it Right the First Time*. Retrieved from http://infrastructureaustralia.gov.au/policy-publications/submissions/published/files/280_associationofconsultingengineersaustralia_SUB.pdf

EOI. (2012). *Sydney Opera House: A Case of Project Management Failure*. Retrieved from www.eoi.es/blogs/cristinagarcia-ochoa/2012/01/14/the-sidney-opera-house-construction-a-case-of-project-management-failure/

Fédération Internationale Des Ingénieurs-Conseils. (2016). *FIDIC Suite of Contracts*. Retrieved from http://fidic.org/bookshop/about-bookshop/which-fidic-contract-should-i-use

Fisher, R. and Ury, W. (1991). *Getting to Yes*. Cornerstone.

Flyvbjerg, B. (2006). *Megaprojects and Risk, An Anatomy of Ambition*. Cambridge University Press.

Flyvbjerg, B. (2014). What You Should Know About Mega Projects And Why: An Overview. *Project Management Journal*, 45 (2): 6–19. DOI: 10.1002/pmj.21409.

Forbes. (2002). *Strategic Alliances*. Retrieved from www.forbes.com/2002/07/18/0719alliance.html

Gladwell, M. (2000). *The Tipping Point*. Little, Brown and Company.

Grattan Institute, Marion Terrill. (2016). *Cost Overruns in Transport Infrastructure*. Retrieved from Grattan Institute: https://grattan.edu.au/report/cost-overruns-in-transport-infrastructure/

Graves, D. C. (2017). *Spiral Dynamics*. Retrieved from https://spiraldynamics.org/2012/06/meet-dr-graves/

Grech, J. (2005). *Creating a Culture of Organisational High Productivity*. Hyder Consulting.

Grech, J. (2006). *The Recruitment Experience*. Hyder Consulting.

Grimsey, D. A. (2017). *Global Developments In Public Infrastructure Procurement*. Edward Elgar.

Haddon-Cave, S. C. (2005). *RAF Nimrod Crash*. Retrieved from The Nimrod Review: www.gov.uk/government/publications/the-nimrod-review

Heinrich, H. (1931). *Industrial Accident Prevention: A Scientific Approach*. McGraw Hill.

Herzberg, F. W. (1959). *The Motivation to Work*. John Wiley and Sons.

HM Government. (2016). *BIM*. Retrieved from www.gov.uk/government/uploads/system/uploads/attachment_data/file/34710/12-1327-building-information-modelling.pdf

Human Synergistics. (2010). *How Culture Works – What Is Culture?* Retrieved from www.humansynergistics.com/resources/content/2016/12/07/how-culture-works

Human Synergistics. (2017). *Human Synergistics Circumplex*. Retrieved from www.humansynergistics.com/about-us/the-circumplex

Independent Project Analysis Inc. (2015). *Engineering News Record*. Retrieved from www.enr.com/articles/2436-owners-take-rap-for-big-project-fails?v=preview

Ingham, J. L. (1955). *The Johari Window, a Graphic Model of Interpersonal Awareness.* University of California.

Institute of Internal Auditors. (2013). *The Three Lines of Defence in Effective Risk Management.* Retrieved from www.iia.org.au/sf_docs/default-source/member-services/thethreelinesofdefenseineffectiveriskmanagementandcontrol_position_paper_jan_2013.pdf

International Standards Organisation. (2012). *ISO 21500:2012 Guidance on Project Management.* Retrieved from www.iso.org/standard/50003.html

International Standards Organisation. (2016). *ISO 31000:2009 Risk Management.* Retrieved from www.iso.org/iso-31000-risk-management.html

Knight, J. (2003). *Big Picture Workshop.* Sinclair Knight & Partners.

Kotler, P. A. (2012). *Marketing Management.* Pearson.

KPMG. (2015). *Global Construction Survey 2015.* Retrieved from https://home.kpmg.com/au/en/home/insights/2016/09/global-construction-survey-building-a-technology-advantage.html

Levinson, J. A. (2004). *Guerrilla Marketing for Consultants: Breakthrough Tactics for Winning Profitable Clients.* John Wiley & Sons.

Lewis, M. (2006). *Blindside: Evolution of the Game.* W W Norton & Company.

Lynch, D. A. (1990). *Strategy of the Dolphin: Scoring a Win in a Chaotic World.* Ballantine Books.

Maister, D. (1993). *Managing the Professional Services Firm.* Simon & Schuster.

Maister, D. A. (2006). *Managing the Multidimensional Organization.* Retrieved from http://davidmaister.com/articles/managing-the-multidimensional-organization/

Maister, D. W. (2001). *The Trusted Advisor.* Simon & Schuster.

McKinsey & Company. (2015a). *The Construction Productivity Imperative.* Retrieved from www.mckinsey.com/industries/capital-projects-and-infrastructure/our-insights/the-construction-productivity-imperative

McKinsey & Company. (2015b). *Managing the People Side of Risk.* Retrieved from www.mckinsey.com/business-functions/risk/our-insights/managing-the-people-side-of-risk.

McKinsey & Company. (2017a). *Enduring Ideas: The Three Horizons of Growth.* Retrieved from www.mckinsey.com/business-functions/strategy-and-corporate-finance/our-insights/enduring-ideas-the-three-horizons-of-growth

McKinsey & Company. (2017b). *Megaprojects: The Good the Bad and the Better.* Retrieved from www.mckinsey.com/industries/capital-projects-and-infrastructure/our-insights/megaprojects-the-good-the-bad-and-the-better?cid=other-eml-alt-mip-mck-oth-1702

McKinsey & Company, Tom Peters. (1979). *Beyond the Matrix Organisation.* Retrieved from www.mckinsey.com/business-functions/organization/our-insights/beyond-the-matrix-organization

Merrell, D. (2016). *Safety Risk.* Retrieved from www.safetyrisk.net/workplace-safety-poems/safety-poems-by-don-merrell/

MinterEllison. (2016). *Construction Law Made Easy.* Retrieved from www.constructionlawmadeeasy.com/roleofasuperintendent

Moore, J. A. (2017). *Great Cities Theme Ten: Program for Engagement.* Retrieved from www.linkedin.com/pulse/great-cities-theme-ten-program-engagement-james-a-moore/?trk=eml-email_feed_ecosystem_digest_01-recommended_articles-7-TixCompany&midToken=AQH3VLAZqQy6Aw&fromEmail=fromEmail&ut=1g29sWTw-nJ7U1

Net Promoter. (2017). *What is Net Promoter?* Retrieved from www.netpromoter.com/know/

Newman, M. (2009). *Emotional Capitalists: The New Leaders.* John Wiley & Sons.

Norton, D. A. (2004). *Strategy Maps*. Harvard Business Review Press.
Oriah. (2017). *Mountain Dreamer*. Retrieved from www.oriahmountaindreamer.com/
Palisade. (2017). *Project Risk Management*. Retrieved from www.palisade.com/projectriskmanagement/
PESTEL Analysis. (2017). Retrieved from www.professionalacademy.com/blogs-and-advice/marketing-theories---pestel-analysis
PRINCE2. (2016). *What Is PRINCE2?* Retrieved from www.prince2.com/aus/what-is-prince2
Product Life Cycle Stages. (2017). Retrieved from http://productlifecyclestages.com/
Project Management Institute. (1996). *PMBOK Guide*. Retrieved from www.pmi.org/pmbok-guide-standards/foundational/pmbok
Project Management Institute. (2013). *A Guide to the Project Management Body of Knowledge*. Project Management Institute.
Project-Management.Com. (2016). *Top 10 Main Causes of Project Failure*. Retrieved from https://project-management.com/top-10-main-causes-of-project-failure/
Project Smart. (2017). Retrieved from www.projectsmart.co.uk/raci-matrix.php
PSMJ. (2016). *Improving the Business Performance of A/E/C Organizations Worldwide*. Retrieved from www.psmj.com/
PwC. (2017). *Client Experience*. Retrieved from www.pwc.com/gx/en/services/advisory/consulting/customer.html
Schein, E. (1985). *Organizational Culture and Leadership*. John Wiley & Sons.
Scott, G. (2013). *Reconsidering High Performance Team Development for Effective Project Delivery*. IPWEA International Public Works Conference.
Siler, A. (2016). *New Field Institute*. Retrieved from www.newfieldinstitute.com.au/
Spielberg, S. (Director). (2015). *Bridge of Spies* [Motion Picture].
Standards Association of Australia. (2006). *AS 4817: Project Performance Measurement using Earned Value*. Retrieved from SAI Global: www.saiglobal.com/pdftemp/previews/osh/as/as4000/4800/4817-2006.pdf
Taleb, N. N. (2007). *The Black Swan: The Impact of the Highly Improbable*. Random House.
The Enneagram Institute. (2017). *The Nine Enneagram Type Descriptions*. Retrieved from www.enneagraminstitute.com/type-descriptions/
The Myers Briggs Foundation. (2017). *Myers Briggs Type Indicators*. Retrieved from www.myersbriggs.org/my-mbti-personality-type/mbti-basics/
Tzu, S. (1957). *The Art of War*. Berkeley Books.
UK Corporate Governance Code. (2017). Retrieved from www.icaew.com/en/library/subject-gateways/corporate-governance/codes-and-reports/uk-corporate-governance-code
Victorian Auditor General's Office. (2012). *Managing Major Projects*. Retrieved from www.audit.vic.gov.au/report/managing-major-projects
Victorian Department of Treasury & Finance. (2006). *Project Alliancing Practitioner's Guide*. Retrieved from https://infrastructure.gov.au/infrastructure/ngpd/files/National_Guide_to_Alliance_Contracting.pdf
Walker, D. (2015). *Collaborative Project Procurement Arrangements*. Project Management Institute.
Walker, J. (2017). *New Zealand Herald*. Retrieved from Special Skills Needed to Manage: www.nzherald.co.nz/business/news/article.cfm?c_id=3&objectid=11836252

Index

Locators in **bold** refer to tables and those in *italics* for figures.